AF477173

Comprehensive
MCQs in
Organic Chemistry

Comprehensive MCQs in Organic Chemistry

Prof. Md. Rageeb Md. Usman

(M. Pharm., FAPP, FICPHS, FSRHCP, FRSH, FSPER)
Joint Secretaries
SPER Central Branch
&
President
IPA/APP/RSH/SRHCP
Maharashtra State Branch
&
Assistant Professor
Smt. S. S. Patil College of Pharmacy,
Chopda, Maharashtra, India

Prof. (Dr.) Sunila T. Patil (M. Pharm. Ph.D.)
Associate Professor
Department of Pharmaceutical Chemistry,
P.S.G.V.P.M's College of Pharmacy, Shahada, (MS)

Prof. (Dr.) R. Y. Chaudhari (M. Pharm. Ph.D.)
Vice-Principal
Hon'ble Loksevak Madhukarrao Chaudhari,
College of Pharmacy, Faizpur, Maharashtra

PharmaMed Press

An imprint of Pharma Book Syndicate
A Unit of BSP Books Pvt. Ltd.
4-4-309/316, Giriraj Lane,
Sultan Bazar, Hyderabad - 500 095.

Comprehensive MCQs in Organic Chemistry by
Md. Rageeb Md. Usman, Sunila T. Patil and R. Y. Chaudhari

© 2018, *by Publisher*

Published by

PharmaMed Press
An imprint of Pharma Book Syndicate

A Unit of BSP Books Pvt. Ltd.
4-4-309/316, Giriraj Lane, Sultan Bazar,
Hyderabad - 500 095.
Phone: 040-23445688, 23445600
Fax: 91+40-23445611
E-mail: info@pharmamedpress.com
www.pharmamedpress.com/pharmamedpress.net

ISBN: 978-93-88305-06-8

Preface

Comprehensive MCQs in Organic Chemistry book intends to provide free learning tools to students who aspire to appear for various entrance examinations. As per PCI new syllabus of pharmacy, the multiple choice questions are compulsory and it targets to undergraduate B. Pharm. students (GATE, GPAT, PET, MH-CET an exam which shapes the life of students) for higher qualification. Thus we have captured several approach-able areas of learning beyond providing students with question bank of their official entrance. This book will facilitate undergraduate and graduate students in assessing their potential and score high in competitive examinations.

It contain the features like excellent coverage of essential topics, contains over 400 multiple choice questions with their answers, updated with new questions from competitive exams and truly accessible to a broad range of students with varying backgrounds.

Due to reliability and objective nature the Multiple Choice Questions (MCQs) are widely used in entrance tests, competitive examination and at some places in certifying examinations. However, MCQ items have been criticized mostly for their lack of relevance to the core or essential learning objectives, because they mostly tend to emphasize on the rare or the unusual. It is therefore of utmost importance to induct MCQ items which test knowledge, understanding and application ability of the student in relevant areas.

Hope this will be helpful....All the best!!!

- Authors

Acknowledgements

The satisfactions, which accomplish a successful completion of any task, are incomplete without the mentioning of the names of those people who makes it possible.

We wish to thank all my colleagues, past & present, without those help this book would have not been written.

We would like to express our utmost gratitude to *Dr. Parloop Bhatt*, HOD of Pharmacology Dept., L. M. College of Pharmacy, Ahmedabad and *Dr. V. R. Patil*, Principal, T. V. E. S College of Pharmacy, Faizpur.

It given us immense pleasure to thank Prof. *Dr. G. P. Vadnere*, Principal, Smt. S. S. Patil College of Pharmacy, Chopda who encouraged me to pursue this attempt in all fronts.

We would express our special thank towards *Dr. S. P. Pawar*, Principal of P. S. G. V. P. M'S College of Pharmacy, Shahada for his encouragement & support.

We wish also to thank our colleagues for helpful comments & suggestions.

We are grateful to our parents for their unconditional love, support and encouragement.

We are also thankful to publishers. We hope this book will leave the desired impression and look forward to receive the comments from the readers.

-Authors

Contents

Contents

About the Authors

Prof. Md. Rageeb Md. Usman Completed B. Pharm., M. Pharm. from North Maharashtra University, Jalgaon; Asst. Prof, at Smt. Sharadchandrika Suresh Patil College of Pharmacy, Chopda. He is highly Professional academician and Researcher in Pharmacy field. Prof. Md. Rageeb over more than 10 Year Experience in Teaching, Research and authorship of books. He has to his Name more than 35 books, 1 chapter in book, 85 Research /Review publications and 205 Presentation/Participation of National/International Conferences. Md. Rageeb Life Member: Asscn. of Pharmaceutical Teachers of India, Soc. of Pharmaceutical Education & Research (Joint Secretary, Central Br), Asscn. of Pharmacy Professionals (President, Maharashtra State Br.), Indian Hospital Pharmacists Asscn, Indian Pharmacy Graduates Asscn, Indian Pharmacists Asscn. (President, Jalgaon Br.), Research Scholar Hub (Maharashtra State Br.), Soc. of Researchers & Healthcare Professionals (President, Maharashtra State Br), Indian Pharmaceutical Asscn; Member: Indian Soc of Pharmacognosy, Asscn of Biotechnology & Pharmacy, International Natural Hygiene Soc; Assoc Editor/Editor of several professional journals and magazines. Prof. Md. Rageeb guided more than 50 students for project. He Conferred Fellowship Award 2013 (twice), 2014, 2015 (twice), Appreciation Award for Poster 2013, Young Performer Award 2013, Young Pharmacy Teacher Award 2014, 2016, 2017, Best Oral Presentation Award 2014, Young Innovative Researcher Award 2014, Appreciation Award for Oral Presentation 2014, Young Talent Award 2014 (twice), 2016, Young Pharmacist Award 2015, Young Excellent Academic Award 2016, Life Time Achievement Award 2016 and Eminent Teacher Award 2017. Md. Rageeb Biography has been included in the renowned directory "Who's Who in the World 2016, "Learned India Educationists Who's Who in the World 2017 and Famous India: Nation's Who's Who 2018.

Dr. Sunila T. Patil working as a Associate Professor M. Pharm., Ph.D (Pharmaceutical Chemistry) & Head of Dept. of Quality Assurance at P.S.G.V.P.M's College of Pharmacy, Shahada. She has completed M. Pharm. from T.V.E.S's College of Pharmacy, Faizpur, affiliated to North Maharashtra University Jalgaon & Ph.D from Jodhpur National University, Jodhpur. She has published Two book as national publication & total 28 national & international research articles in impact factor journals. Total 20 state, national & international seminar/conferences attended & total 10 papers presented in various national & international conferences. She got one research grant under VCRMS, NMU Jalgaon. She has been nominated as Student Welfare Officer (2013-2017) from North Maharashtra University Jalgaon. She has successfully organized 7 one day workshops & 1 national level seminar sponsored by North Maharashtra University Jalgaon conducted at P.S.G.V.P.M's college of pharmacy Shahada. She has awarded as Best Programme Officer by Dr. Sudhir Meshram, vice-chancellor of North Maharashtra University Jalgaon for achieving best University rank in Maharashtra state at RTM University Nagpur (Avhan-2014). She is a Life member of APTI & SPER. She has appointed as university representative & invited as Resource person in staff development programme sponsored by M.S.B.T.E. She is a coordinator of Anti-ragging committee. Also holding PG, Alumni & Cultural In-charge of P.S.G.V.P.M's college of pharmacy Shahada & successfully organized all activities in Institute.

Prof. (Dr.) R Y Chaudhari is Vice-Principal and Head of the Department, Pharmaceutical Chemistry at Honorable Loksevak Madhukarrao Chaudhari College of Pharmacy, Faizpur, Maharashtra, affiliated to North Maharashtra University, Jalgaon has received bachelor and master degree in Pharmacy from Poona College of Pharmacy Pune in 1992 and 1994 respectively and received PhD degree (2007) in Pharmaceutical Sciences from University Department of Pharmaceutical Sciences RTM Nagpur University Nagpur. He is authoring of a neumorous Professional Publications and a reference book Reterosynthetic Analysis and Synthesis of Drugs: The Disconnection Approach, Nirali Prakashan, Pune. He also sevres on the Board of Studies North Maharashtra University Jalgaon. He has guided about 60 M. Pharm students and 7 Ph.D. students.

<table><tr><td>CHAPTER
1</td><td># ATOMS, MOLECULES AND CHEMICAL BONDING</td></tr></table>

1. **Which is the ground-state electronic configuration of chlorine?**

   (a) $1s^2 2s^2 2p^5$ (b) $1s^2 2s^2 2p^6 3s^2 3p^4$

   (c) $1s^2 2s^2 2p^6 3s^2 3p^5$ (d) $1s^2 2s^2 2p^8 3s^2 3p^5$

2. **Which is the ground-state electronic configuration of Na^+?**

   (a) $1s^2 2s^2 2p^8$ (b) $1s^2 2s^2 2p^6 3s^1$

   (c) $1s^2 2s^2 2p^6 3s^2$ (d) $1s^2 2s^2 2p^6$

3. **Which of the following atoms or ions has the electronic configuration $1s^2 2s^2 2p^6 3s^2 3p^6$?**

   (a) Ne   (b) K   (c) Cl^- (d) Mg^{2+}

4. **Which of the following elements is less electro negative than carbon?**

   (a) Si   (b) N   (c) Cl   (d) F

5. **Which of the following bonds is the most polar?**

   (a) C-H   (b) O-H   (c) C-O   (d) N-H

6. **Which is a correct Lewis structure for carbon dioxide, CO_2?**

   (a) $\ddot{O}=C=\ddot{O}$ (b) $\ddot{:}O=C=O\ddot{:}$

   (c) $:O=\ddot{C}=O:$ (d) $:\ddot{O}-C\equiv O:$

7. Which is a correct Lewis structure for carbonic acid, H_2CO_3?

(a)

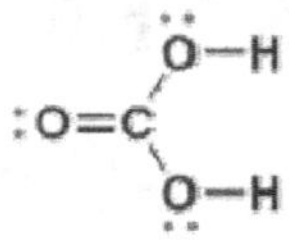

(b)

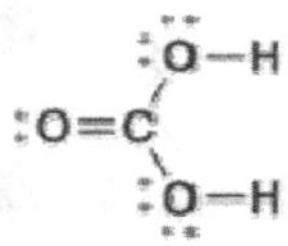

(c)

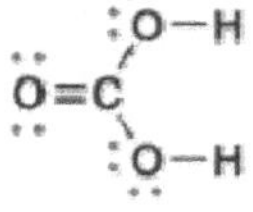

(d)

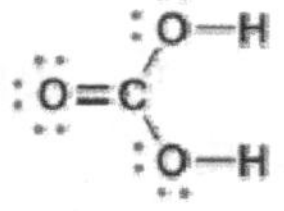

8. Which is a correct Lewis structure for hydrogen carbonate ion, HCO_3^-?

(a)

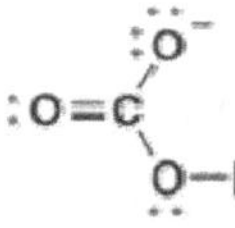

(b)

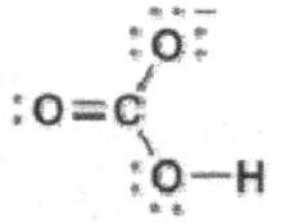

(c)

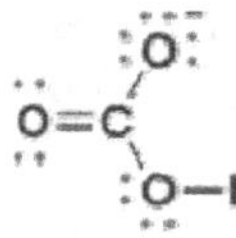

(d)

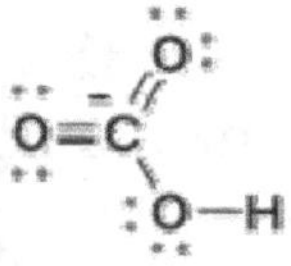

9. Which is a correct Lewis structure for hydrogen cyanide, HCN?

(a)    H—C≡N

(b)   H—C≡N:

(c)   H—C≡N:

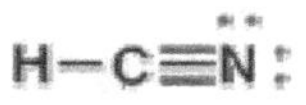

(d)    H—C≡N:

10. Which is a correct Lewis structure for nitric acid, HNO_3?

(a) 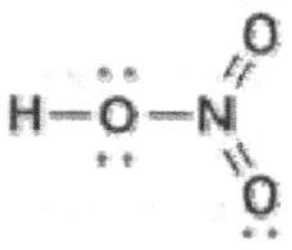   H—O—N

(b) 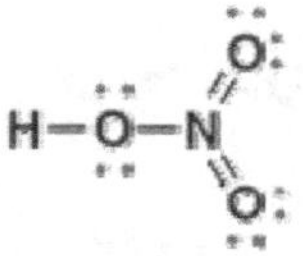   H—O—N

(c) 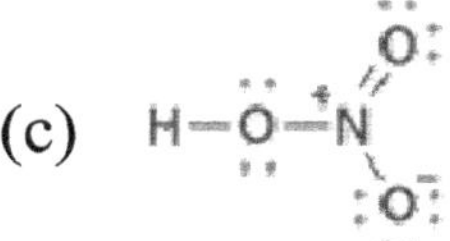   (d) 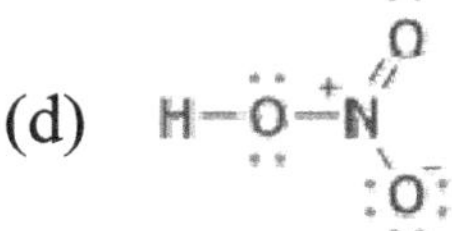

11. Which of the following does not have the ground-state configuration $1s^2 2s^2 2p^6$?

    (a) Ne   (b) Na^+ (c) Cl^- (d) F^-

12. Which of the following elements is the most electronegative?

    (a) B   (b) C   (c) Cl   (d) N

13. Which of the following elements is the most electropositive?

    (a) B   (b) C   (c) Cl   (d) N

14. Which of the following elements is the most electronegative?

    (a) Br   (b) Cl   (c) F   (d) I

15. Which of the following elements is the most electropositive?

    (a) Br   (b) Cl   (c) F   (d) I

16. Which of the following compounds has an ionic bond?

    (a) H_2O (b) NH_4Cl (c) CH_3Cl (d) CH_3Li

17. Which of the following molecules does not have a dipole moment?

    (a) CH_3Cl (b) CH_2Cl_2 (c) $CHCl_3$ (d) CCl_4

18. Which of the following contains an atom (other than hydrogen) which lacks an octet of valence electrons?

    (a) NH_3 (b) H_3O^+ (c) BH_3 (d) NH_4^+

19. Which of the following is a correct Lewis structure of diazomethane, CH_2N_2?

    (a) 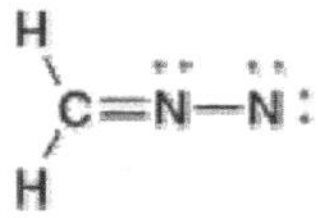 $H_2C=\ddot{N}-\ddot{\ddot{N}}:$

    (b) 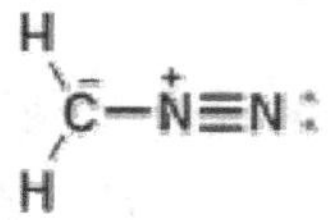 $H_2C-\overset{+}{N}\equiv N:$

    (c) 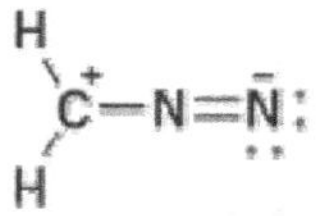 $H_2\overset{+}{C}-N=\ddot{\ddot{N}}:$

    (d) 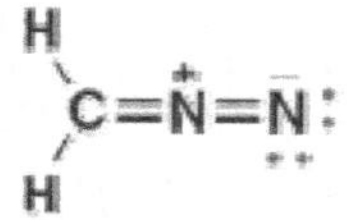 $H_2C=\overset{+}{N}=\ddot{N}:$

20. Which of the following Lewis structures of protonated methanamide is incomplete?

    (a)

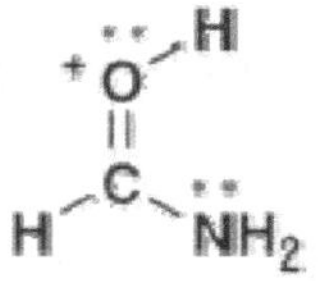

    (b)

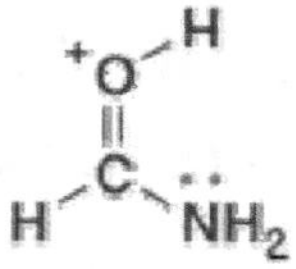

    (c)

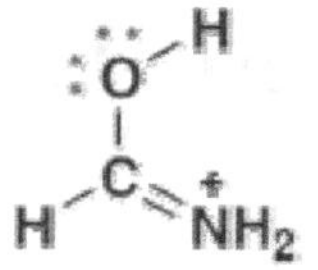

    (d)

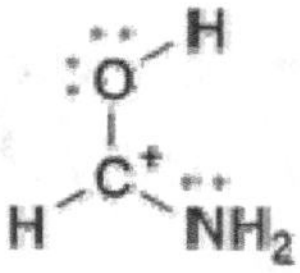

KEYS

1. (c)	2. (d)	3. (c)	4. (a)	5. (b)
6. (a)	7. (c)	8. (c)	9. (d)	10. (c)
11. (c)	12. (c)	13. (a)	14. (c)	15. (d)
16. (b)	17. (d)	18. (c)	19. (d)	20. (b)

MOLECULAR STRUCTURE AND SHAPES OF ORGANIC MOLECULES

1. Which set of approximate bond angles at C1, C2, and N of the following molecule indicates the correct shape?

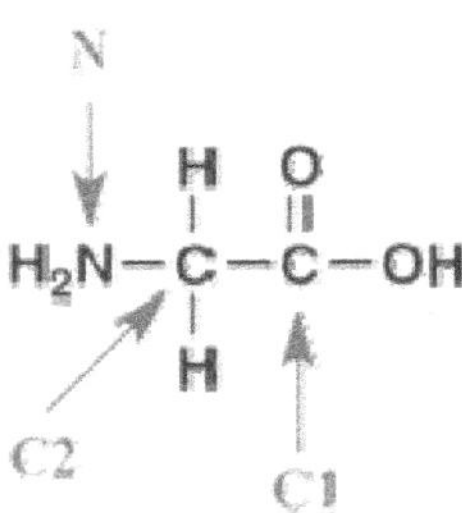

(a) C1 120°, C2 120°, N 120°

(b) C1 109.5°, C2 120°, N 109.5°

(c) C1 109.5°, C2 120°, N 120°

(d) C1 120°, C2 109.5°, N 109.5°

2. Which set of approximate bond angles at C1, C2, and N of the following molecule indicates the correct shape?

(a) C1 120°, C2 120°, N 120°

(b) C1 120°, C2 120°, N 109.5°

(c) C1 109.5°, C2 120°, N 120°

(d) C1 120°, C2 109.5°, N 109.5°

3. **Which set of approximate bond angles at C1, O, and C2 of the following molecule indicates the correct shape?**

    (a) C1 120°, O 120°, C2 120°

    (b) C1 120°, O 120°, C2 109.5°

    (c) C1 120°, O 109.5°, C2 120°

    (d) C1 120°, O 109.5°, C2 109.5°

4. **Which set of hybridization states of C1, C2, and N of the following molecule is correct?**

    (a) C1 sp^2, C2 sp^3, N sp^3

    (b) C1 sp^3, C2 sp^2, N sp^3

    (c) C1 sp^3, C2 sp^2, N sp^2

    (d) C1 sp^2, C2 sp^2, N sp^3

5. **Which set of hybridization states of C1, C2, and N of the following molecule is correct?**

    (a) C1 sp^2, C2 sp^3, N sp^3

    (b) C1 sp^2, C2 sp^2, N sp^3

    (c) C1 sp^2, C2 sp^2, N sp^2

    (d) C1 sp^2, C2 sp^3, N sp^2

6. Which species does not contain an sp^3- hybridized atom?

   (a) BH_3 (b) BH_4^- (c) NH_3 (d) NH_4^+

7. Which species contains an sp^3-hybridized atom?

   (a) H_3C^+ (b) $H_2C=\overset{+}{O}H$

   (c) H_3O^+ (d) $H_2C=NH$

8. Which is the correct molecular formula of the following molecule?

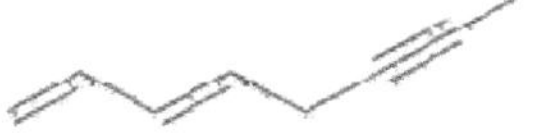

   (a) C_8H_8 (b) C_8H_{10}

   (c) C_7H_8 (d) C_7H_{10}

9. Which is the correct molecular formula of the following molecule?

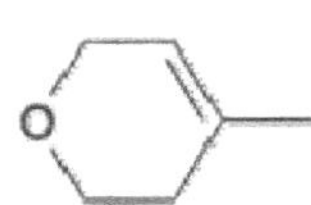

   (a) $C_7H_{10}O$

   (b) $C_7H_{12}O$

   (c) C_6H_8O

   (d) $C_6H_{10}O$

10. Which is the correct molecular formula of the following molecule?

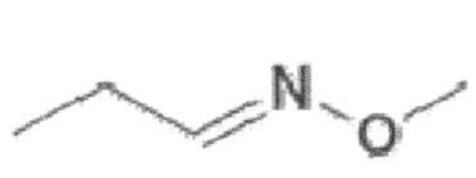

   (a) C_3H_7NO

   (b) C_3H_8NO

   (c) C_4H_9NO

   (d) $C_4H_{10}NO$

11. **Which set of approximate bond angles at C1, C2, and C3 of the following molecule indicates the correct shape?**

C1 C2 C3

$N \equiv C - CH = CH_2$

   (a) C1 120°, C2 120°, C3 120°
   (b) C1 180°, C2 120°, C3 109.5°
   (c) C1 180°, C2 120°, C3 120°
   (d) C1 180°, C2 180°, C3 120°

12. **Which set of approximate bond angles at C, the central O, and N of the following molecule indicates the correct shape?**

$CH_3 - O - NO_2$

   (a) C 109.5°, O 109.5°, N 120°
   (b) C 109.5°, O 120°, N 120°
   (c) C 120°, O 120°, N 120°
   (d) C 109.5°, O 120°, N 109.5°

13. **Which set of hybridization states of C1, C2, and C3 of the following molecule is correct?**

$HO - \overset{\overset{O}{\|}}{C} - CH = C = CH_2$

C1 C2 C3

   (a) sp^2, sp^2, sp^2
   (b) sp^2, sp^2, sp
   (c) sp^3, sp^2, sp
   (d) sp^3, sp^2, sp^2

14. **Which molecule contains an sp-hybridized atom?**

   (a) HCO_2H (b) HNO_3 (c) HNO_2 (d) HCN

15. **Which species contains an sp^2-hybridized atom?**

   (a) BeH_2 (b) BH_3 (c) NH_3 (d) H_3O^+

16. Which of the following structures represents a molecule different from the others?

(a) 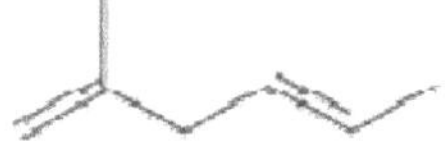

(b)

(c)

(d)

17. Which is the correct molecular formula of the following diene?

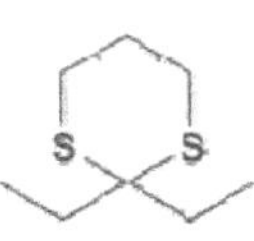

(a) C_7H_{14} (b) C_7H_{12}

(c) C_6H_{12} (d) C_6H_{10}

18. Which is the correct molecular formula of the following dithiane?

(a) $C_8H_{16}S_2$

(b) $C_8H_{18}S_2$

(c) $C_6H_{12}S_2$

(d) $C_6H_{14}S_2$

19. Which alkene has the (*E*)-configuration?

(a)

(b)

(c)

(d)

20. Which of the following statements is wrong?

(a) When two orbitals overlap in-phase with each other, a bonding molecular orbital forms.

(b) When two orbitals overlap out-of-phase with each other, an ant bonding molecular orbital forms.

(c) When one of two atoms connected by a σ bond rotates about the bond axis, orbital overlap is lost.

(d) When one of two atoms connected by a π bond rotates about the bond axis, orbital overlap is lost.

KEYS

1. (d)	**2.** (a)	**3.** (d)	**4.** (a)	**5.** (c)
6. (a)	**7.** (c)	**8.** (b)	**9.** (d)	**10.** (c)
11. (c)	**12.** (a)	**13.** (b)	**14.** (d)	**15.** (b)
16. (c)	**17.** (b)	**18.** (a)	**19.** (d)	**20.** (c)

CHAPTER 3

ORGANIC COMPOUNDS
(Functional Groups, Intermolecular Interactions and Physical Properties)

1. Which compound is not a constitutional isomer of but-2-ene, $CH_3CH=CHCH_3$?

(a) 

(b) 

(c) 

(d)

2. Which compound is different from compound 1?

(a) $CH_3COCH_2CH_3$

(b) $CH_3CH_2\overset{\overset{\displaystyle O}{\|}}{C}CH_2CH_3$

(c) $CH_3CH_2COCH_3$

(d) $CH_3CH_2\overset{\overset{\displaystyle O}{\|}}{C}CH_3$

3. Which compound is not ether?

(a) $CH_3CH_2OCH_2CH_3$

(b) $CH_3COCH_2CH_3$

(c) 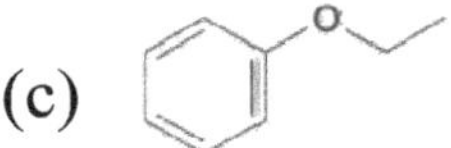

(d) 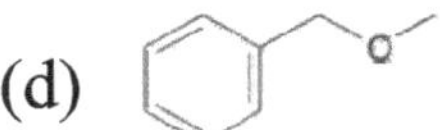

4. **Which compound is not a carboxylic acid?**

   (a) $CH_3CH_2CO_2H$

   (b)

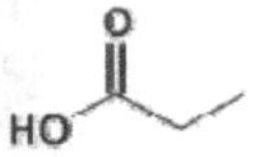

   (c)

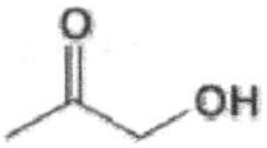

   (d)

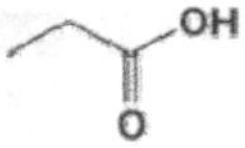

5. **Which compound is a tertiary amine?**

   (a) $(CH_3)_3CNH_2$

   (b) $CH_3CH_2\underset{\underset{CH_3}{|}}{\overset{\overset{CH_3}{|}}{C}}NH_2$

   (c) $CH_3CH_2\underset{\overset{|}{CH_3}}{N}CH_3$

   (d) $(CH_3CH_2)_2NH$

6. **Which compound is a tertiary alcohol?**

   (a) $(CH_3)_3COH$

   (b) $CH_3CH_2\underset{\overset{|}{OH}}{C}HCH_3$

   (c) $CH_3CH_2\overset{\overset{H}{|}}{O}CH_3$

   (d) $(CH_3CH_2)_2CHOH$

7. **Which compound has the highest boiling point?**

   (a) $CH_3CH_2CH_2CH_3$

   (b) $CH_3CH_2\overset{\overset{O}{||}}{C}H$

   (c) $CH_3CH_2OCH_3$

   (d) $CH_3CH_2CH_2OH$

8. **Which compound is least soluble in water?**

   (a) $CH_3CH_2CH_2CH_3$

   (b) $CH_3CH_2\overset{\overset{O}{||}}{C}H$

   (c) $CH_3CH_2OCH_3$

   (d) $CH_3CH_2CH_2OH$

9. Which of (a)-(d) is the correct IUPAC name of the following compound?

(a) 4-ethylpent-3-ene

(b) 2-ethylpent-2-ene

(c) 3-methylhex-3-ene

(d) 4-methylhex-3-ene

10. Which of (a)-(d) is the correct IUPAC name of the following compound?

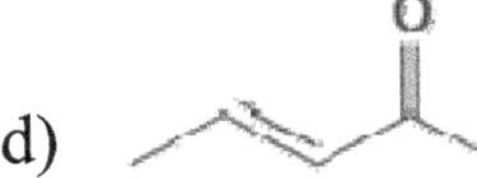

(a) 2-chloro-3-methylpent-3-ene

(b) 4-chloro-3-methylpent-2-ene

(c) 3-methyl-4-chloropent-2-ene

(d) 3-methyl-2-chloropent-3-ene

11. Which compound is an aldehyde?

(a)   (b)

(c)   (d)

12. Which compound is different from the others?

(a) methyl ethyl ketone   (b) pentan-2-one

(c) 2-pentanone   (d) methyl propyl ketone

13. Which compound is a secondary alcohol?

(a) butan-1-ol   (b) butan-2-ol

(c) Isobutyl alcohol   (d) 2-methylpropan-2-ol

14. Which compound is a secondary amine?

(a) $(CH_3CH_2)_2CHNH_2$

(b) [structure: cyclohexane ring with NH]

(c) [structure: cyclohexyl–NH₂]

(d) [structure: pyridine ring with N]

15. Which compound has the highest boiling point?

(a) butan-1-ol

(b) butanal

(c) butanone

(d) butanoic acid

16. Which compound is most soluble in water?

(a) butan-1-ol

(b) butanal

(c) butanone

(d) butanoic acid

17. Which alcohol has the highest boiling point?

(a) butan-1-ol

(b) butan-2-ol

(c) isobutyl alcohol

(d) *t*-butyl alcohol

18. Which alcohol is most soluble in water?

(a) butan-1-ol

(b) butan-2-ol

(c) isobutyl alcohol

(d) *t*-butyl alcohol

19. Which of (a)-(d) is the correct IUPAC name of the following compound?

$$CH_3CH_2\underset{\underset{Br}{|}}{C}HCH=\underset{\underset{CH_2CH_3}{|}}{C}CH_2CH_3$$

(a) 3-ethyl-5-bromohept-3-ene

(b) 5-bromo-3-ethylhept-3-ene

(c) 3-bromo-5-ethylhept-4-ene

(d) 1,1-diethyl-3-bromopent-1-ene

20. **Which of (a)-(d) is the correct IUPAC name of the following compound?**

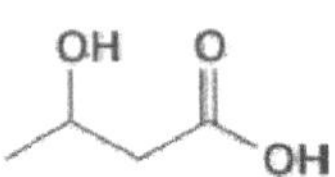

    (a) 2-hydroxybutanoic acid
    (b) 3-hydroxybutanoic acid
    (c) 2-hydroxypropanoic acid
    (d) 1-carboxypropan-2-ol

KEYS

1. (d)	**2.** (b)	**3.** (b)	**4.** (c)	**5.** (c)
6. (a)	**7.** (d)	**8.** (a)	**9.** (c)	**10.** (b)
11. (c)	**12.** (a)	**13.** (b)	**14.** (b)	**15.** (d)
16. (d)	**17.** (a)	**18.** (d)	**19.** (b)	**20.** (b)

CONFORMATION AND STRAIN IN MOLECULES

1. Which structure is of a compound different from the following?

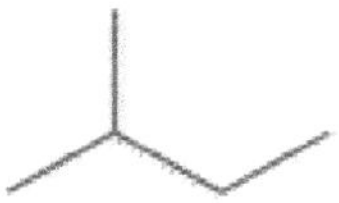

(a)

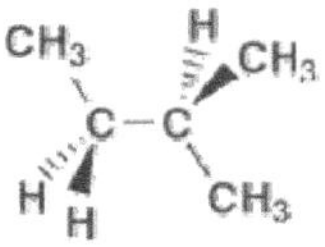

(b)

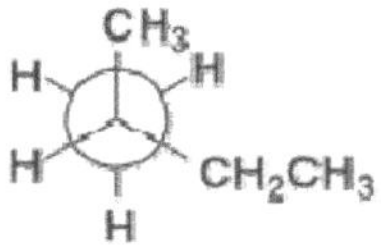

(c)

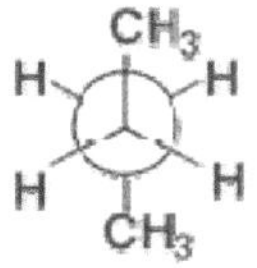

(d)

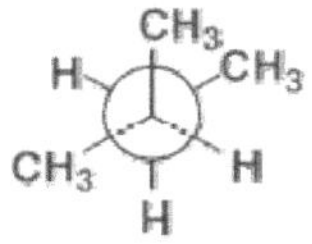

2. Which structure is of a compound different from the following?

$(CH_3)_2CHCHCH_2CH_3$ with CH_3 branch

(a)

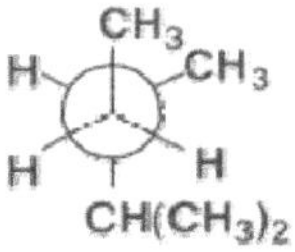

(b)

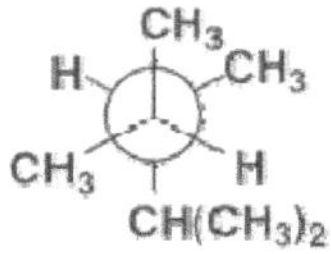

(c)

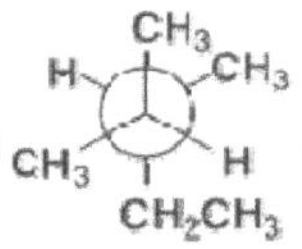

(d)

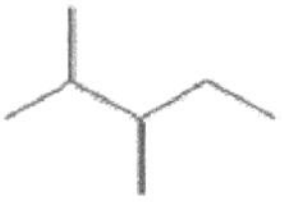

3. **Which of (a)-(d) is the most stable conformation?**

(a) 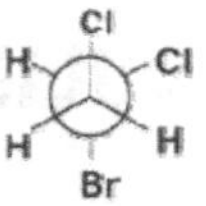   (b)

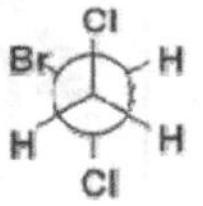

(c) 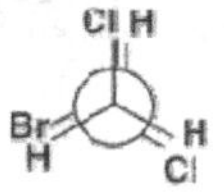   (d)

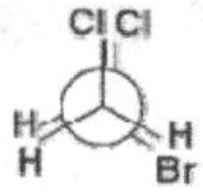

4. **Which of (a)-(d) is the most unstable conformation?**

(a) 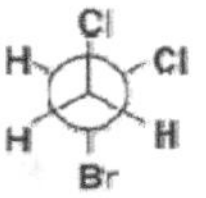   (b)

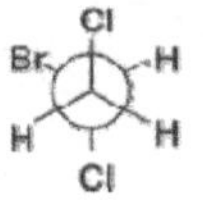

(c) 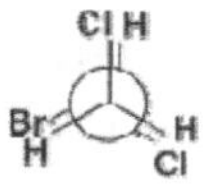   (d)

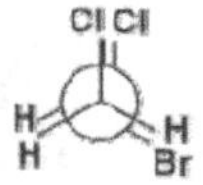

5. **Which of (a)-(d) is the most stable conformation?**

(a) 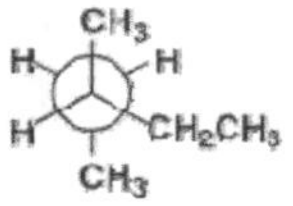   (b)

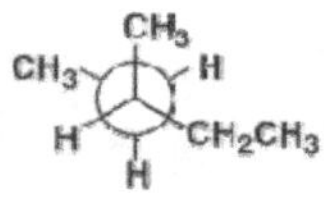

(c) 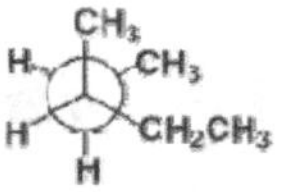   (d)

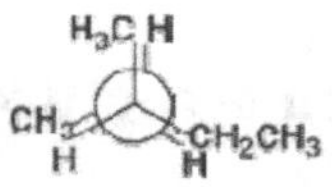

6. **Which of (a)-(d) is the most unstable conformation?**

(a) 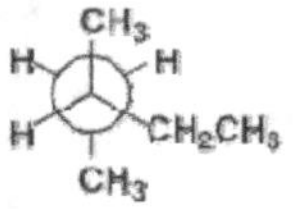   (b)

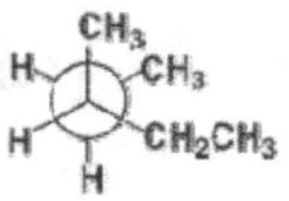

(c) 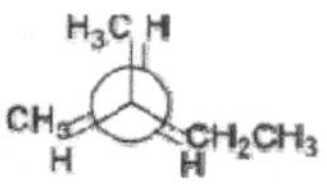   (d)

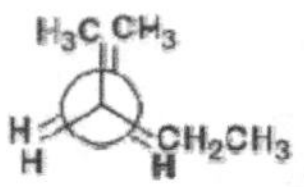

7. **Which is the most stable structure of 1-isopropyl-4-methylcyclohexane?**

(a) 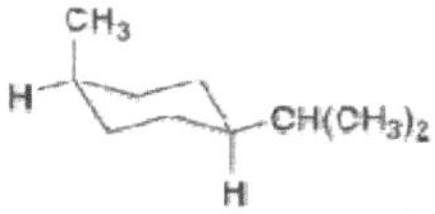 CH_3 — H — $CH(CH_3)_2$ — H

(b) 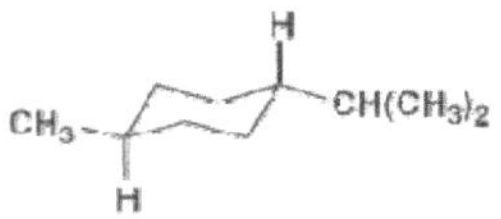 H — CH_3 — $CH(CH_3)_2$ — H

(c) 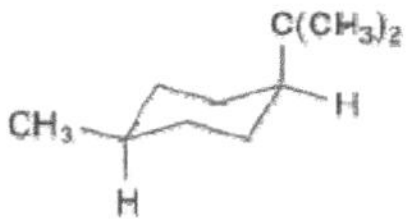 $C(CH_3)_2$ — CH_3 — H — H

(d) 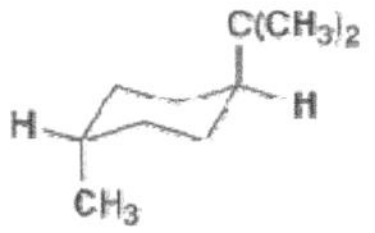 $C(CH_3)_2$ — H — H — CH_3

8. **Which structure is different from the following?**

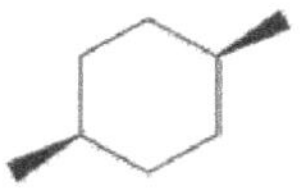

(a) 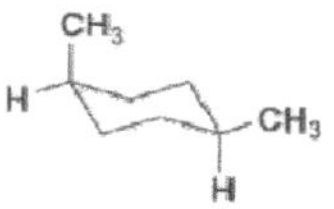 CH_3 — H — CH_3 — H

(b) 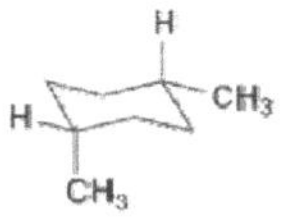 H — H — CH_3 — CH_3

(c) 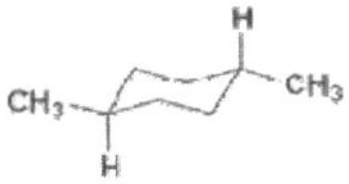 H — CH_3 — CH_3 — H

(d) 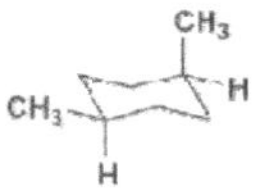 CH_3 — CH_3 — H — H

9. **Which structure is different from the following?**

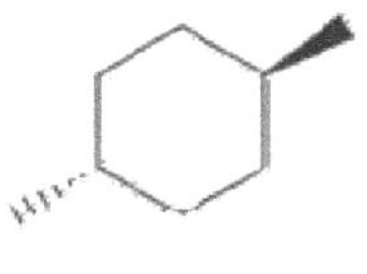

(a) 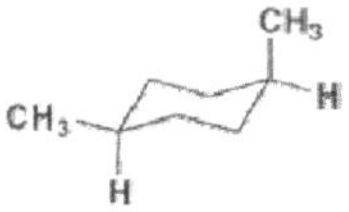 CH_3 — CH_3 — H — H

(b) 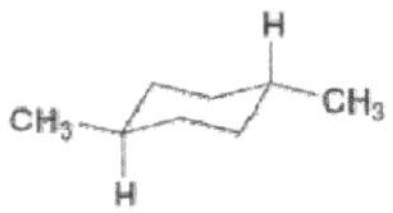 H — CH_3 — CH_3 — H

(c) 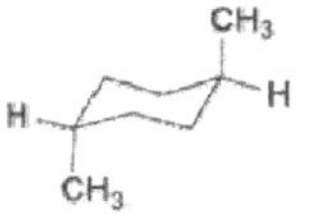 CH_3 — H — H — CH_3

(d) 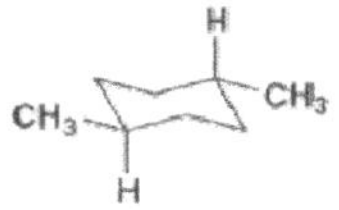 H — CH_3 — CH_3 — H

10. **Which structure is different from the following?**

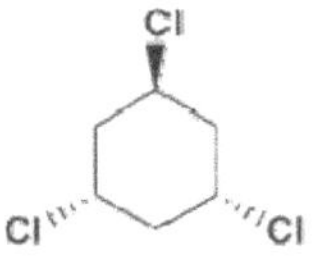

(a) 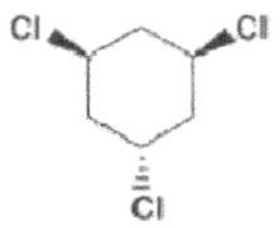 Cl — Cl — Cl

(b) 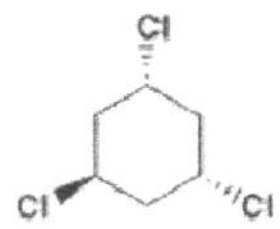 Cl — Cl — Cl

(c) 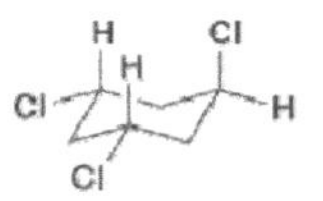 H — H — Cl — Cl — H — Cl

(d) 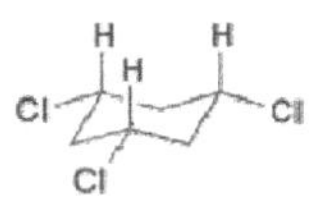 H — H — Cl — Cl — Cl — H

11. Which compound is different from the others?

(a)

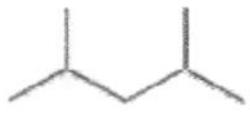

(b)

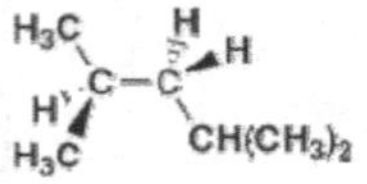

(c)

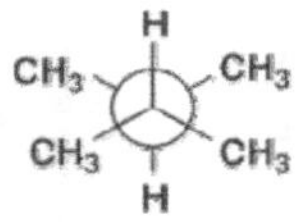

(d)

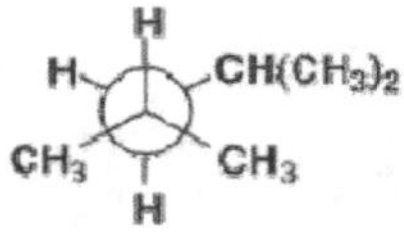

12. Which compound is different from the others?

(a)

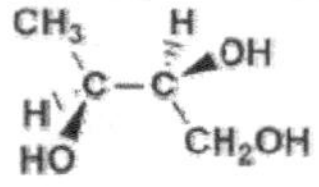

(b)

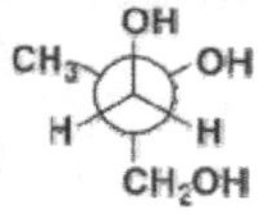

(c)

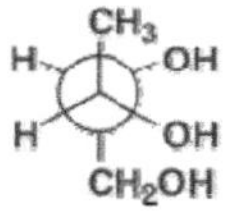

(d)

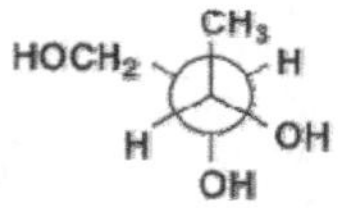

13. Which is the most stable conformation?

(a)

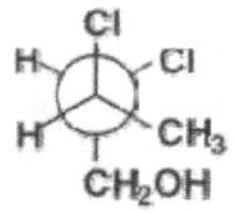

(b)

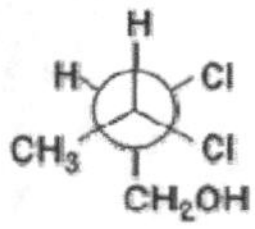

(c)

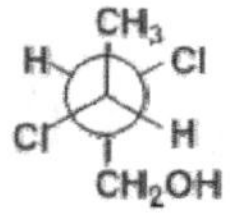

(d)

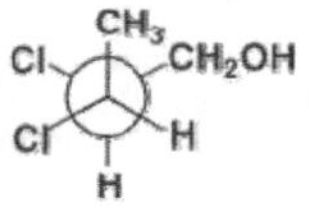

14. Which of (a)–(d) is the most stable?

(a)

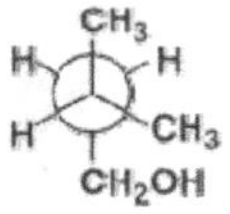

(b)

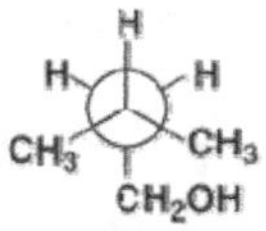

(c)

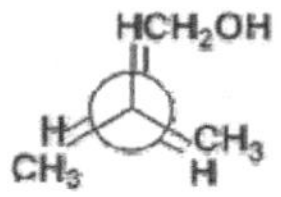

(d)

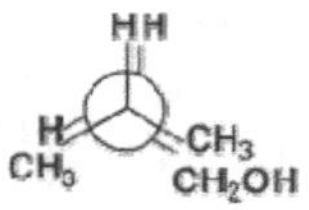

15. **Which of (a)-(d) is the most unstable?**

(a) 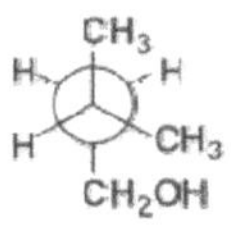 [Newman projection: front carbon with CH_3 (top), H and H; back carbon with H, CH_3; CH_2OH at bottom]

(b) 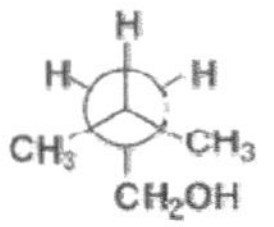 [Newman projection: front carbon with H (top), H and H; back carbon with CH_3, CH_3; CH_2OH at bottom]

(c) 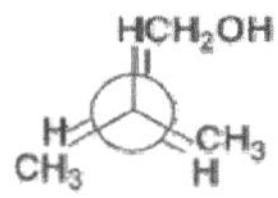 [Newman projection: HCH_2OH top; H, CH_3; CH_3, H]

(d) 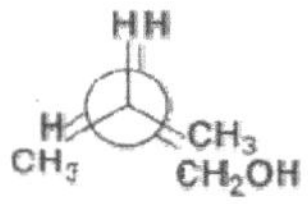 [Newman projection: H H top; H, CH_3; CH_3, CH_2OH]

16. **Which structure is different from the following?**

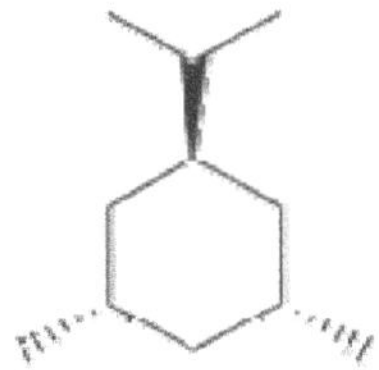

(a) 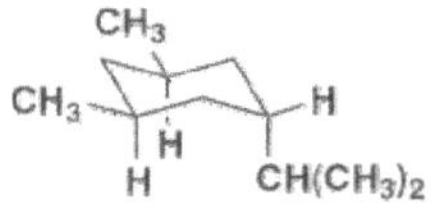 [cyclohexane chair: CH_3, CH_3, H, H, H, $CH(CH_3)_2$]

(b) 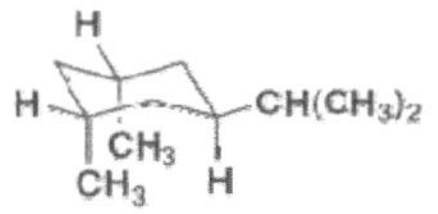 [cyclohexane chair: H, H, $CH(CH_3)_2$, CH_3, CH_3, H]

(c) 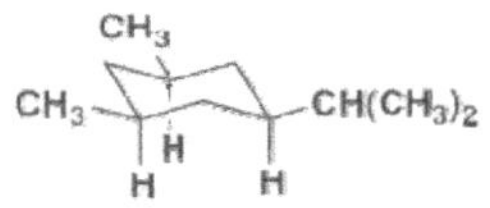 [cyclohexane chair: CH_3, CH_3, H, $CH(CH_3)_2$, H, H]

(d) 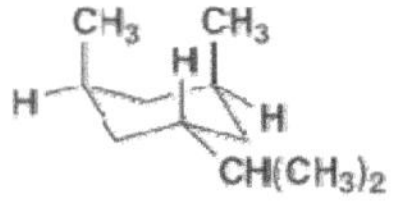 [cyclohexane chair: CH_3, CH_3, H, H, H, $CH(CH_3)_2$]

17. **Which structure is different from the following?**

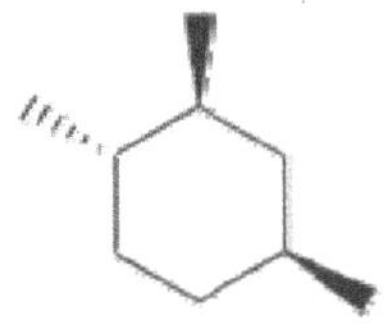

(a) 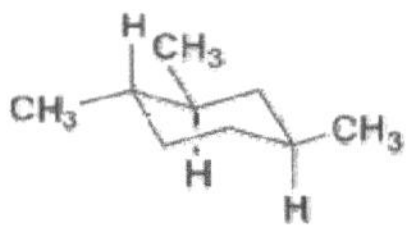 [cyclohexane chair: H, CH_3, CH_3, CH_3, H, H]

(b) 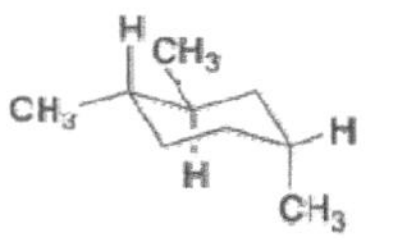 [cyclohexane chair: H, CH_3, CH_3, H, H, CH_3]

(c)

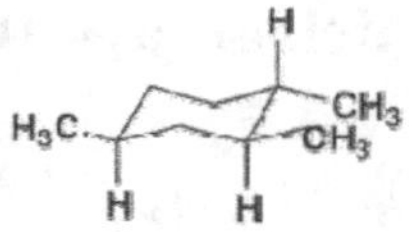

(d)

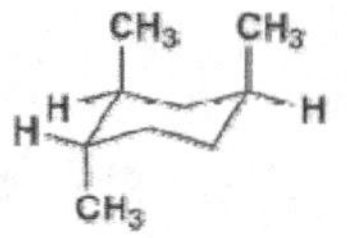

18. Which conformation is most unstable?

(a)

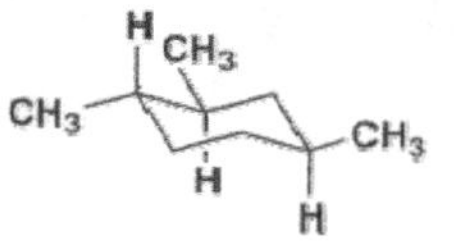

(b)

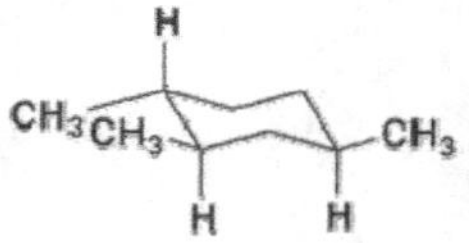

(c)

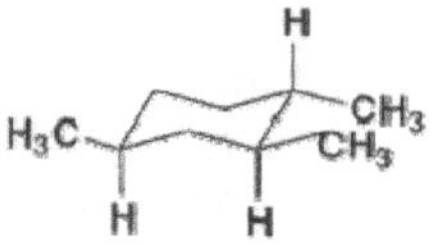

(d)

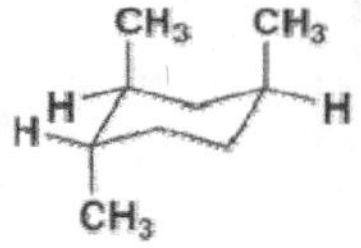

19. Which of the following statements regarding cycloalkanes is wrong?

(a) Any disubstituted cycloalkane can have cis-trans isomers.

(b) The planar form of any cycloalkane with a ring larger than cyclopropane will not be the most stable conformation.

(c) Cyclopentane is nonplanar to avoid the torsional strain between adjacent C-H bonds.

(d) The least strained form of any unsubstituted cyclo alkane is the chair conformation of cyclo hexane.

20. Which of the following statements regarding chair cyclohexane is wrong?

(a) The dihedral angle of the two axial bonds on adjacent carbons is 180°.

(b) The dihedral angle of the two equatorial bonds onadjacent carbons is 60°.

(c) The dihedral angle between the axial bond and the equatorial bond on adjacent carbons is 120°.

(d) The axial hydrogen atoms on C1, C3, and C5 form an equilateral triangle (as do C1, C3, and C5 themselves and the equatorial hydrogens on them).

KEYS

1. (c)	**2.** (b)	**3.** (a)	**4.** (d)	**5.** (b)
6. (d)	**7.** (b)	**8.** (c)	**9.** (a)	**10.** (d)
11. (c)	**12.** (b)	**13.** (c)	**14.** (a)	**15.** (d)
16. (c)	**17.** (b)	**18.** (d)	**19.** (a)	**20.** (c)

1. **Which compound has a conjugated system?**

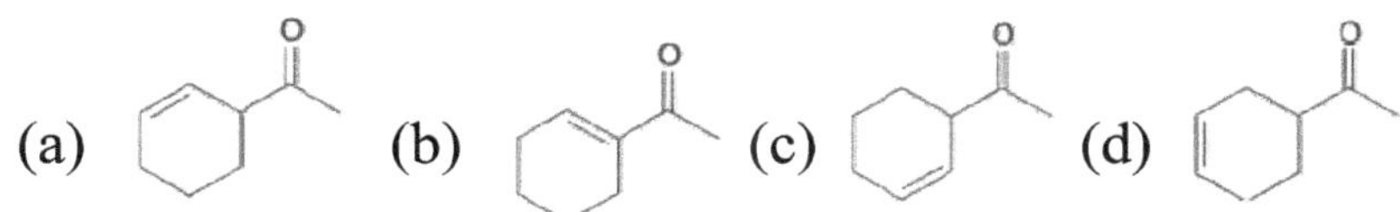

(a)   (b)   (c)   (d)

2. **Which compound has a conjugated system?**

(a) 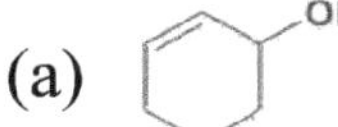   (b) 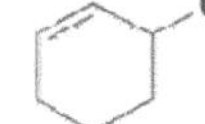   (c)    (d) 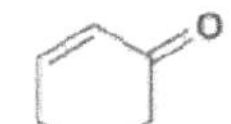

3. **Which is a defective resonance contributor of the following cation?**

(a)   (b)

(c)   (d)

4. **Which is a defective resonance contributor of naphthalene?**

(a)    (b) 

(c) 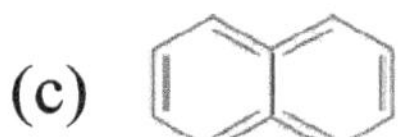   (d) 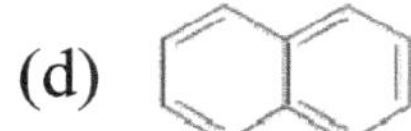

5. **Which of (a)-(d) is the most important resonance contributor for the following?**

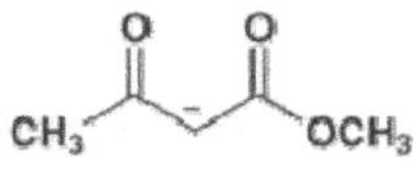

(a) 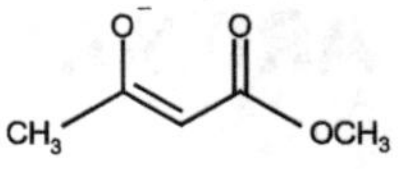

(b) 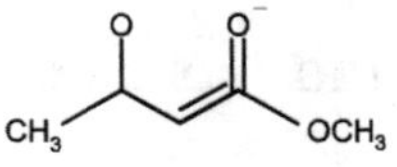

(c) 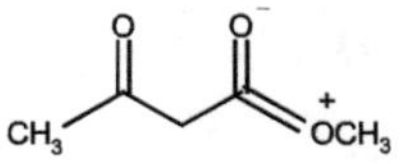

(d) 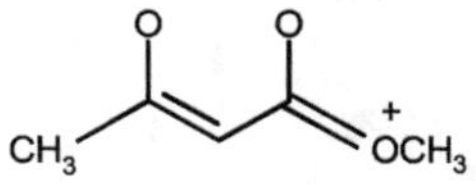

6. **Which of (a)-(d) is the most important resonance contributor for the following?**

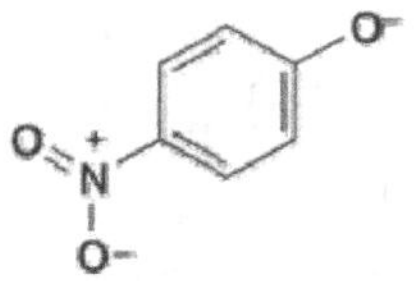

(a) 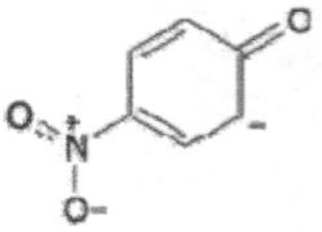

(b) 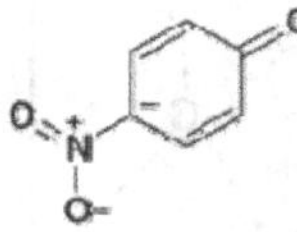

(c) 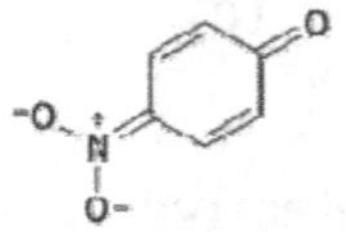

(d) 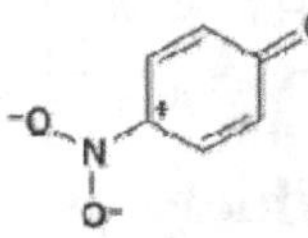

7. **Which of (a)-(d) is not aromatic?**

(a) (b) (c) (d)

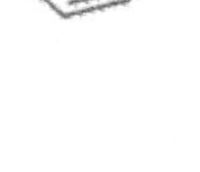

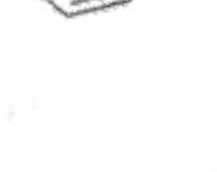

8. **Which compound is not aromatic?**

(a)   (b)

(c)   (d)

9. **Which of (a)-(d) is not aromatic?**

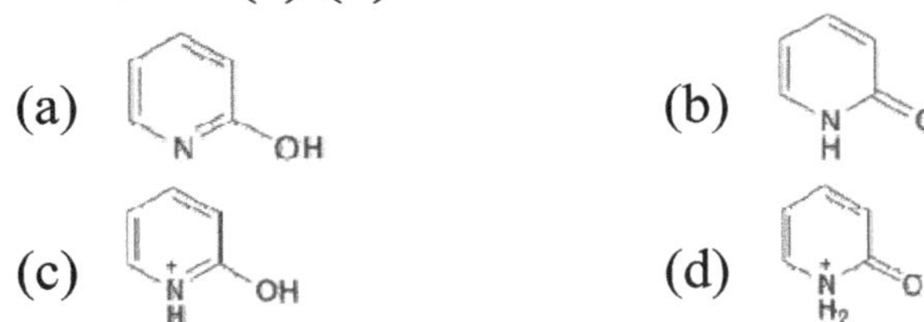

(a)   (b)

(c)   (d)

10. **Which of the following statements is false?**

(a) The π molecular orbitals of buta-1,3-diene are derived from 2p atomic orbitals of carbon atoms, and there are four of them.

(b) Any orbital can accommodate up to two electrons, so buta-1, 3-diene has eight π electrons.

(c) The complete set of molecular orbitals obtained by combining atomic orbitals includes an increased number of nodes.

(d) Some orbitals do not have any nodes.

11. Which compound does not have a conjugated system?

    (a) [structure with OH]

    (b) [structure]

    (c) [structure]

    (d) [structure]

12. Which compound does not have a conjugated system?

    (a) $CH_2{=}CH{-}C{\equiv}C{-}CH_3$

    (b) $CH_3{-}CH{=}C{=}CH{-}CH_3$

    (c) $CH_3{-}CH{=}CH{-}OCH_3$

    (d) $CH_3{-}CH{=}CH{-}CN$

13. Which compound does not have a conjugated system?

    (a) [structure with CH_3, CH_2, CH_2^-]

    (b) [structure: $CH_3{-}C({=}O){-}O^-$]

    (c) [structure: $HO{-}\overset{+}{N}({=}O){-}O^-$]

    (d) [structure: $CH_3{-}C(H)(OH){-}CN$]

14. Which of the following is a defective resonance contributor of *p*-acetylphenol?

    (a) [structure]

    (b) [structure]

    (c) [structure]

    (d) [structure]

15. Which of the following is a defective resonance contributor of *p*-nitrophenol?

(a) 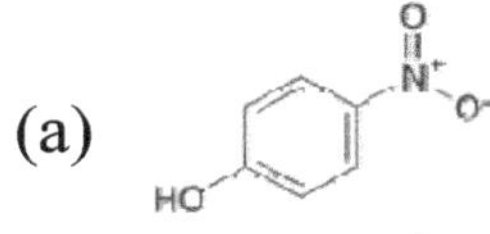

(b) 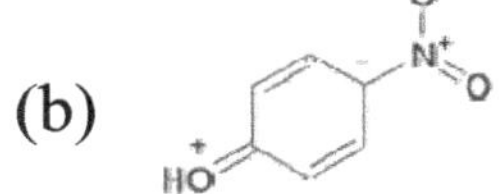

(c) (d)

16. Which of the following is a defective resonance contributor of the acetamidinium ion?

(a) (b)

(c) (d)

17. Which is the least important resonance contributor of methyl propenoate?

(a) (b)

(c) 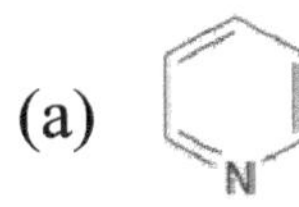

(d) 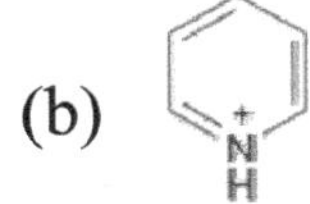

18. Which of the following is not aromatic?

(a) (b)

(c) (d)

19. Which compound is not aromatic?

(a)   (b)

(c)   (d)

20. Which of the following is not an allyl anion analogue?

(a) $CH_3CH=CH_2$ (b) $^-CH_2CO_2Et$

(c) $^-CH_2NO_2$ (d) $CH_3OCH=CH_2$

KEYS

1. (b)	**2.** (d)	**3.** (d)	**4.** (c)	**5.** (a)
6. (c)	**7.** (b)	**8.** (c)	**9.** (d)	**10.** (b)
11. (c)	**12.** (b)	**13.** (d)	**14.** (b)	**15.** (d)
16. (c)	**17.** (b)	**18.** (d)	**19.** (b)	**20.** (a)

CHAPTER **6**   **ACIDS AND BASES**

1. **Which of the following pairs does not show an acid and its conjugate base?**

   (a) HNO_3 and NO_3^- (b) H_2SO_4 and HSO_4^-

   (c) H_2SO_4 and SO_4^{2-} (d) HSO_4^- and SO_4^{2-}

2. **Which of the following pairs does not show an acid and its conjugate base?**

   (a) $H_3\overset{+}{N}CH_2CO_2H$ and $H_3\overset{+}{N}CH_2CO_2^-$

   (b) $H_2NCH_2CO_2H$ and $H_3\overset{+}{N}CH_2CO_2^-$

   (c) $H_2NCH_2CO_2H$ and $H_2NCH_2CO_2^-$

   (d) $H_3\overset{+}{N}CH_2CO_2^-$ and $H_2NCH_2CO_2^-$

3. **Which compound is most acidic?**

   (a) CH_4 (b) NH_3

   (c) H_2O (d) H_2S

4. **Which compound is least acidic?**

   (a) CH_4 (b) NH_3

   (c) H_2O (d) H_2S

5. Which compound is most acidic?

(a) FCH_2CO_2H (b) $ClCH_2CO_2H$

(c) $BrCH_2CO_2H$ (d) ICH_2CO_2H

6 Which compound is least acidic?

(a) FCH_2CO_2H (b) $ClCH_2CO_2H$

(c) $BrCH_2CO_2H$ (d) ICH_2CO_2H

7. Which compound is most basic?

(a) 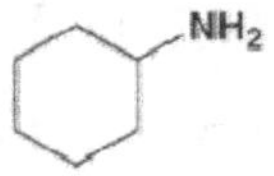

(b) 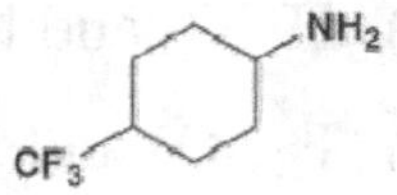

(c) 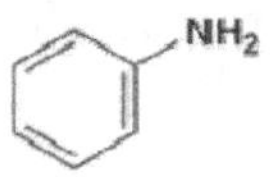

(d) 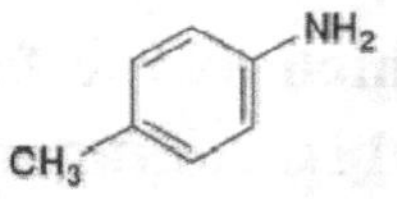

8. Which compound is least basic?

(a) 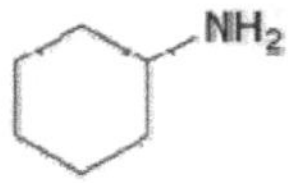

(b) 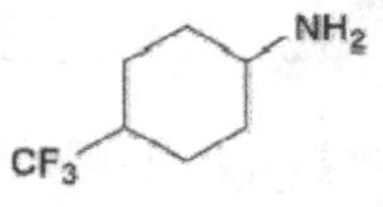

(c) 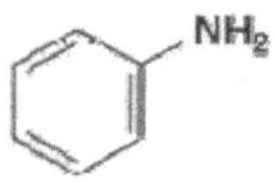

(d) 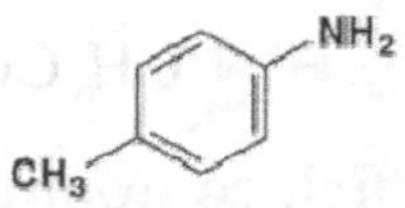

9. Which compound is most acidic?

(a) 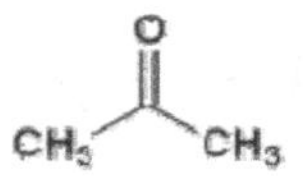

(b) 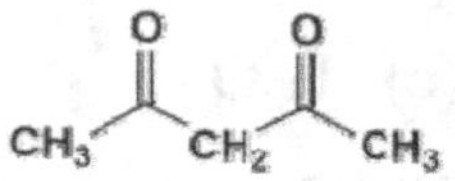

(c) 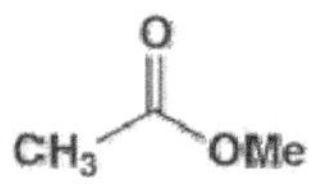

(d) 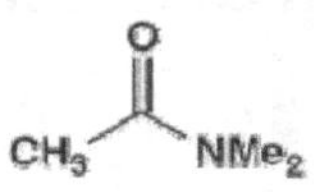

10. **Which compound is least acidic?**

(a) CH_3—CO—CH_3

(b) CH_3—CO—CH_2—CO—CH_3

(c) CH_3—CO—OMe

(d) CH_3—CO—NMe_2

11. **Which of the following pairs does not show an acid and its conjugate base?**

(a) NH_3 and NH_2^-

(b) NH_4^+ and NH_3

(c) NH_4^+ and NH_2^-

(d) H_3PO_4 and $H_2PO_4^-$

12. **Which carboxylic acid is most acidic?**

(a) $CH_3CH_2CH_2CO_2H$

(b) $CH_3CH_2CH(Cl)CO_2H$

(c) $CH_3CH(Cl)CH_2CO_2H$

(d) $ClCH_2CH_2CH_2CO_2H$

13. **Which substituted benzoic acid is most acidic?**

(a) 3-Cl-benzene-CO_2H

(b) 4-Cl-benzene-CO_2H

(c) 3-CH_3-benzene-CO_2H

(d) 4-CH_3-benzene-CO_2H

14. **Which substituted benzoic acid is least acidic?**

(a) 3-Cl-benzene-CO_2H

(b) 4-Cl-benzene-CO_2H

(c) 3-CH_3-benzene-CO_2H

(d) 4-CH_3-benzene-CO_2H

15. Which substituted phenol is most acidic?

(a) 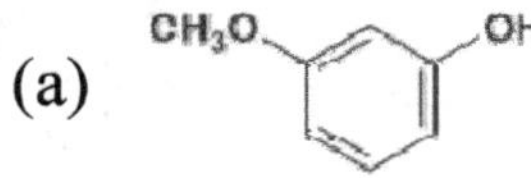

(b)

(c)

(d)

16. Which substituted phenol is least acidic?

(a)

(b)

(c)

(d)

17. Which substituted aniline is most basic?

(a)

(b)

(c)

(d)

18. Which substituted aniline is least basic?

(a)

(b)

(c)

(d)

19. **Which of (a)-(d) shows the increasing order of acidity of alcohols 1-4?**

   (a) $1 < 2 < 3 < 4$ (b) $1 < 2 < 4 < 3$

   (c) $1 < 3 < 2 < 4$ (d) $4 < 3 < 2 < 1$

20. **Which of ((a)-((d) shows the increasing order of basicity of compounds 1-4?**

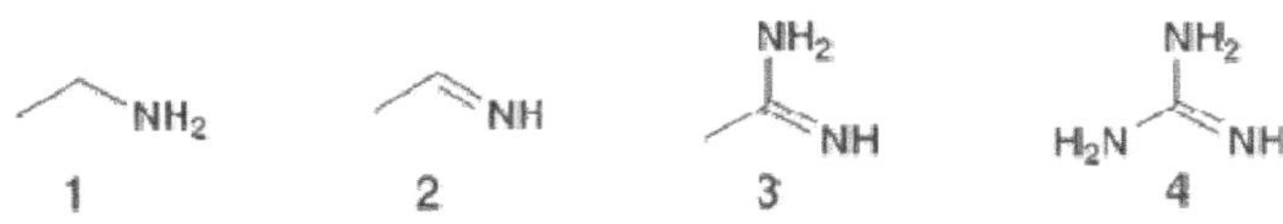

   (a) $1 < 2 < 3 < 4$ (b) $1 < 2 < 4 < 3$

   (c) $2 < 1 < 3 < 4$ (d) $4 < 3 < 2 < 1$

KEYS

1. (c)	2. (b)	3. (d)	4. (a)	5. (a)
6. (d)	7. (a)	8. (c)	9. (b)	10. (d)
11. (c)	12. (b)	13. (a)	14. (d)	15. (d)
16. (b)	17. (a)	18. (d)	19. (b)	20. (c)

ORGANIC REACTIONS AND CONCEPT OF MECHANISM

1. Which of the following cannot react as a nucleophile?

   (a) CH_3NH_2 (b) $(CH_3)_2NH$

   (c) $(CH_3)_3N$ (d) $(CH_3)_4N^+$

2. Which of the following cannot react as a nucleophile?

   (a) CH_3OH (b) CH_3O^-

   (c) $CH_3O^+H_2$ (d) CH_3OCH_3

3. Which of the following cannot react as a nucleophile?

   (a) $H2C=CH2$ (b) $BH3$

   (c) $H2C=NH$ (d) $CH3CH2SH$

4. Which of the following is not a typical electrophile?

   (a) Cl_2 (b) HBr

   (c) $(CH_3)_4N^+$ (d) Br_2

5. Which of the following reactions is an elimination?

   (a) $(CH_3)_3CCl \longrightarrow (CH_3)_2C=CH_2 + HCl$

   (b) $CH_3\overset{O}{\overset{\|}{C}}OH + NH_3 \longrightarrow CH_3\overset{O}{\overset{\|}{C}}O^- + NH_4^+$

   (c) $H_2C=O + H_2O \longrightarrow H_2C(OH)_2$

(d) $CH_3\overset{O}{\overset{\|}{C}}OH$ + CH_3OH $\longrightarrow$ $CH_3\overset{O}{\overset{\|}{C}}OCH_3$ + H_2O

6. Which of the following reactions is a substitution?

(a) $(CH_3)_3CCl$ $\longrightarrow$ $(CH_3)_2C{=}CH_2$ + HCl

(b) $CH_3\overset{O}{\overset{\|}{C}}OH$ + NH_3 $\longrightarrow$ $CH_3\overset{O}{\overset{\|}{C}}O^-$ + NH_4^+

(c) $H_2C{=}O$ + H_2O $\longrightarrow$ $H_2C(OH)_2$

(d) $CH_3\overset{O}{\overset{\|}{C}}OH$ + CH_3OH $\longrightarrow$ $CH_3\overset{O}{\overset{\|}{C}}OCH_3$ + H_2O

7. Which of the following reactions is an addition?

(a) $(CH_3)_3CCl$ $\longrightarrow$ $(CH_3)_2C{=}CH_2$ + HCl

(b) $CH_3\overset{O}{\overset{\|}{C}}OH$ + NH_3 $\longrightarrow$ $CH_3\overset{O}{\overset{\|}{C}}O^-$ + NH_4^+

(c) $H_2C{=}O$ + H_2O $\longrightarrow$ $H_2C(OH)_2$

(d) $CH_3\overset{O}{\overset{\|}{C}}OH$ + CH_3OH $\longrightarrow$ $CH_3\overset{O}{\overset{\|}{C}}OCH_3$ + H_2O

8. Which of the statements (a)-(d) about the reaction profile below is false?

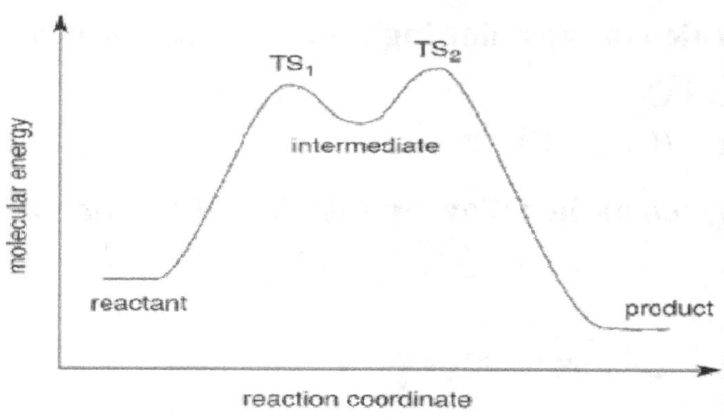

    (a) The product is more stable than the reactant.

    (b) The first step is rate determining.

    (c) The reaction is exothermic.

    (d) The equilibrium constant is > 1 if the molar entropy change is negligible.

9. The following stepwise reactions comprise a mechan-ism for the acid-catalyzed hydrolysis of an ester. Which step of (a)-(d) is an addition?

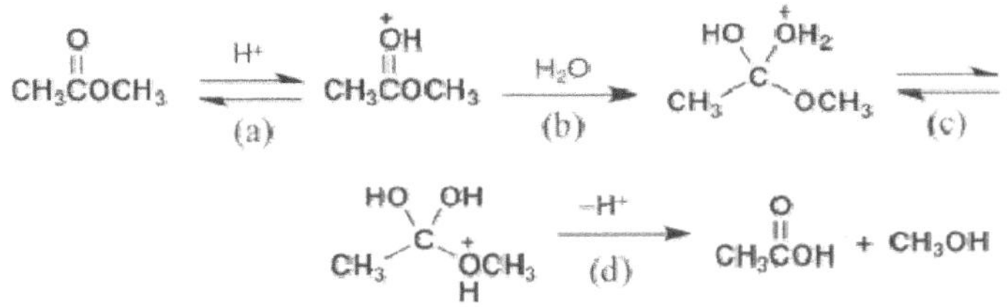

    (a) Step (a)   (b) Step (b)

    (c) Step (c)   (d) Step (d)

10. The following stepwise reactions comprise a mechanism for the acid-catalysed hydrolysis of an ester. Which step of (a)-(d) is an elimination?

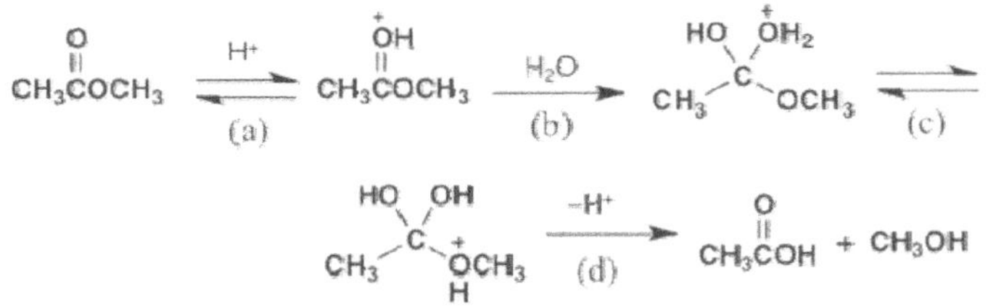

    (a) Step (a)   (b) Step (b)

    (c) Step (c)   (d) Step (d)

11. Which of the following can react as a nucleophile?

    (a) $(CH_3)_3B$ (b) $(CH_3)_3CH$

    (c) $(CH_3)_3N$ (d) $(CH_3)_3O^+$

12. **Which of the following does not normally react as a nucleophile?**

    (a) $(CH_3)_2O$

    (b) $(CH_3)_2OH^+$

    (c) CH_3CH_2OH

    (d) $CH_3CH_2O^-$

13. **Which of the following reacts most readily as an electrophile?**

    (a) HBr

    (b) CH_3Br

    (c) CH_3OH

    (d) $(CH_3)_4N^+$

14. **Which of the following normally reacts as an electrophile?**

    (a) $(CH_3)_3CH$

    (b) $H_2C{=}CH_2$

    (c) $(CH_3)_2O$

    (d) H_3Cl

15. **Which of the following can react readily either as a nucleophile or as an electrophile?**

    (a) $(CH_3)_3B$

    (b) $(CH_3)_2O$

    (c) CH_3CO_2H

    (d) $H_2C{=}CH_2$

16. **Which of the statements ((a)-((d) about the following reaction profile is wrong?**

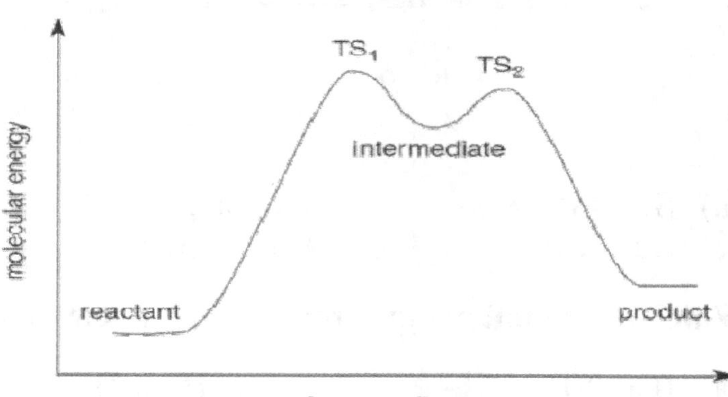

(a) The rate-determining step of this reaction is the first step.

(b) This reaction is exothermic.

(c) The product is less stable than the reactant.

(d) Equilibrium is in favour of the reactant if the molar entropy change is negligible.

17. Which of the following statements is wrong?

(a) The higher the energy of its HOMO, the more reactive a nucleophile will usually be.

(b) The higher the energy of its LUMO, the more reactive an electrophile will usually be.

(c) In an exothermic reaction, the enthalpy of the product(s) is lower than that of the   reactant(s).

(d) The equilibrium constant of an endergonic reaction is smaller than 1.

18. Which of the following statements is wrong?

(a) It is not necessary for a nucleophile to have an unshared electron pair.

(b) A species can react as an electrophile if it contains an atom (other than hydrogen) with an incomplete valence octet.

(c) A species can react as an electrophile, even if it has one or more unshared electron pairs.

(d) Any species bearing a lone pair can normally react as a nucleophile.

19. **Where is the use of a curly arrow wrong in the following reaction sequence? Specify by (a)-(d).**

(a)   (b)   (c)

(d)

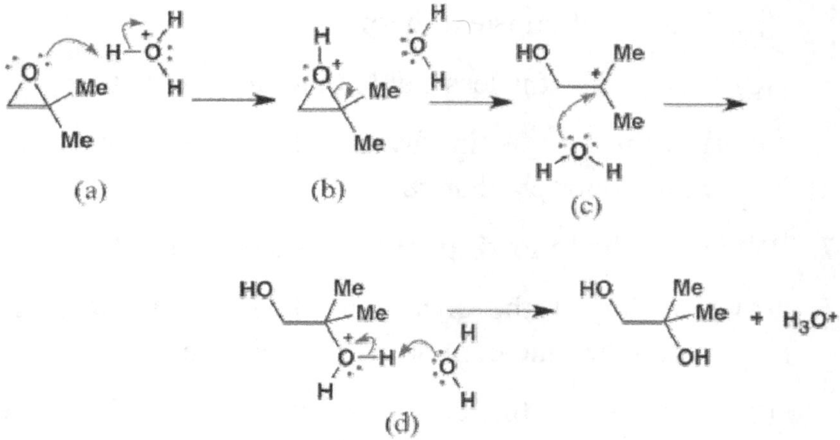

(a) Stage (a)   (b) Stage (b)

(c) Stage (c)   (d) Stage (d)

20. **Where is the use of a curly arrow wrong in the following reaction sequence? Specify by (a)-(d).**

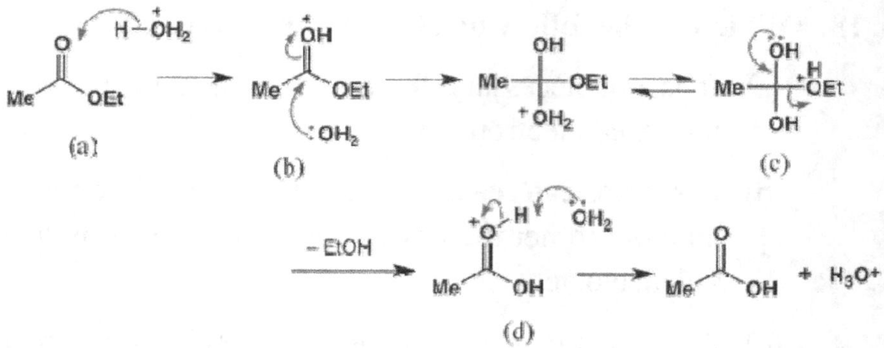

(a)   (b)   (c)

(d)

(a) Stage (a)   (b) Stage (b)

(c) Stage (c)   (d) Stage (d)

<u>**KEYS**</u>

1. (d)	**2.** (c)	**3.** (b)	**4.** (c)	**5.** (a)
6. (d)	**7.** (c)	**8.** (b)	**9.** (b)	**10.** (d)
11. (c)	**12.** (b)	**13.** (a)	**14.** (d)	**15.** (c)
16. (b)	**17.** (b)	**18.** (d)	**19.** (b)	**20.** (a)

NUCLEOPHILLIC ADDITION REACTIONS

1. **Which of the following has the largest equilibrium constant for the formation of a cyanohydrin?**

   (a) CH_3CHO (b) CH_3COCH_3 (c) CH_3CH_2CHO (d) C_6H_5CHO

2. **Which of the following has the largest equilibrium constant for hydration?**

   (a) CH_3CHO (b) $ClCH_2CHO$ (c) CH_3OCH_2CHO (d) Cl_2CHCHO

3. **Which of the following has the largest equilibrium constant for hydration?**

   (a) C_6H_5CHO (b) CH_3-C_6H_4CHO

   (c) O_2N-C_6H_4CHO (d) CH_3O-C_6H_4CHO

4. **Which of the following has the smallest equilibrium constant for hydration?**

   (a) C_6H_5CHO (b) CH_3-C_6H_4CHO

(c) 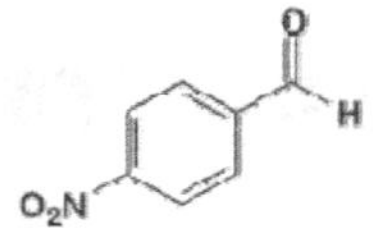 [structure: 4-nitrobenzaldehyde, O₂N]

(d) 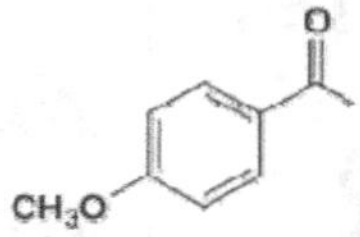 [structure: 4-methoxybenzaldehyde, CH₃O]

5. **Which is normally the main product when a mixture of aldehyde RCHO and an excess of alcohol R'OH is treated with an acid catalyst?**

(a) 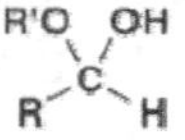 [structure: R'O OH / C / R H]

(b) [structure: R'O O⁻ / C / R H]

(c) [structure: R'O OR' / C / R H]

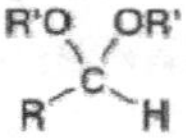

(d) 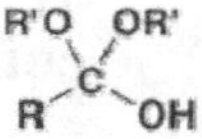 [structure: R'O OR' / C / R OH]

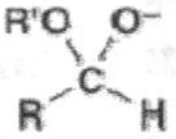

6. **Which is normally the main product when a mixture of aldehyde RCHO and an excess of alcohol R'OH is treated with a base catalyst?**

(a) 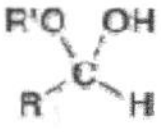 [structure: R'O OH / C / R H]

(b) [structure: R'O O⁻ / C / R H]

(c) [structure: R'O OR' / C / R H]

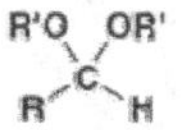

(d) 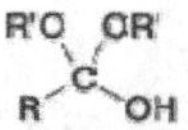 [structure: R'O OR' / C / R OH]

7. **Which of ((a)-((d) is least likely as an intermediate or product in the following reaction?**

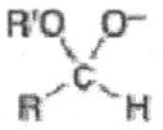

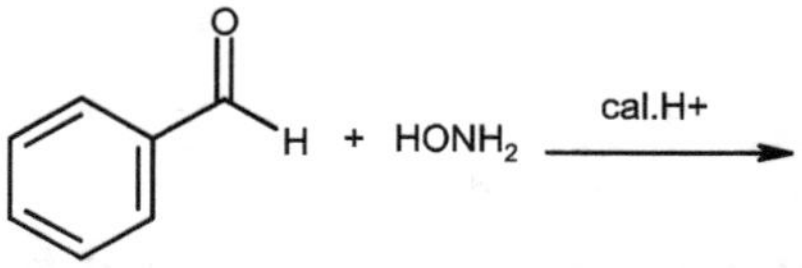

[structure: benzaldehyde + HONH₂ → (cal.H+)]

(a) 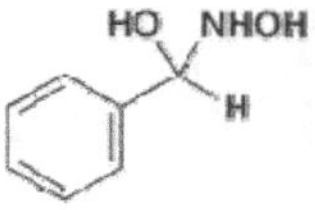 [structure: HO NHOH / C6H5 / H]

(b) 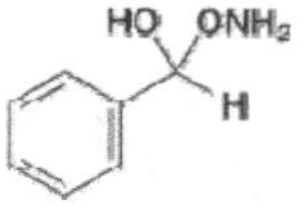 [structure: HO ONH₂ / C6H5 / H]

(c) 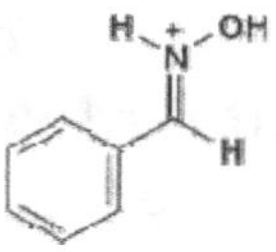 [structure: H–N⁺–OH / =CH–C6H5 / H]

(d) 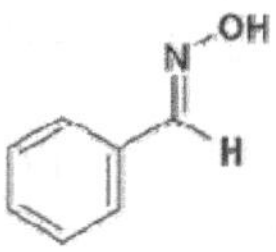 [structure: N–OH / =CH–C6H5 / H]

8. Which of the following reactions does not give the compound indicated as a main product?

(a) 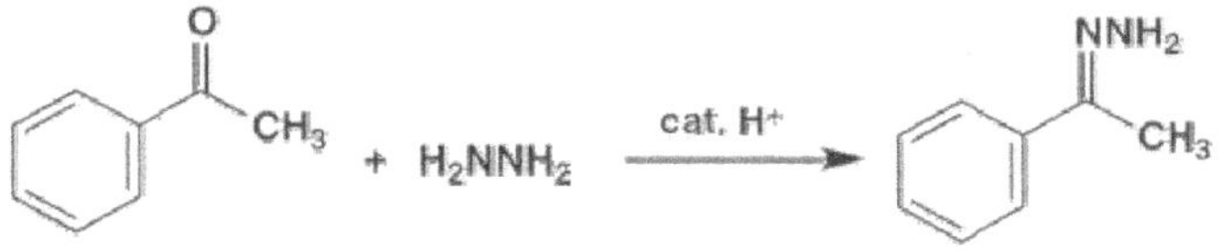 [structure: acetophenone] $+$ H_2NNH_2 $\xrightarrow{\text{cat. } H^+}$ [structure: hydrazone, NNH$_2$]

(b) 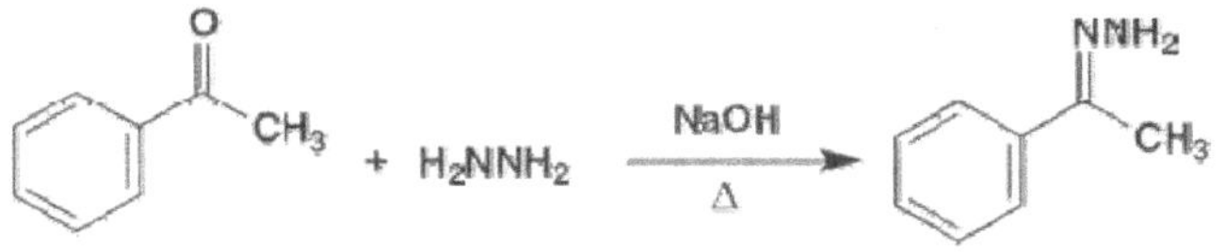 [structure: acetophenone] $+$ H_2NNH_2 $\xrightarrow[\Delta]{\text{NaOH}}$ [structure: hydrazone, NNH$_2$]

(c) CH_3CH_2Br $+$ PPh_3 $\longrightarrow$ $CH_3\overset{+}{C}H_2\text{–}PPh_3\ Br^-$

(d) 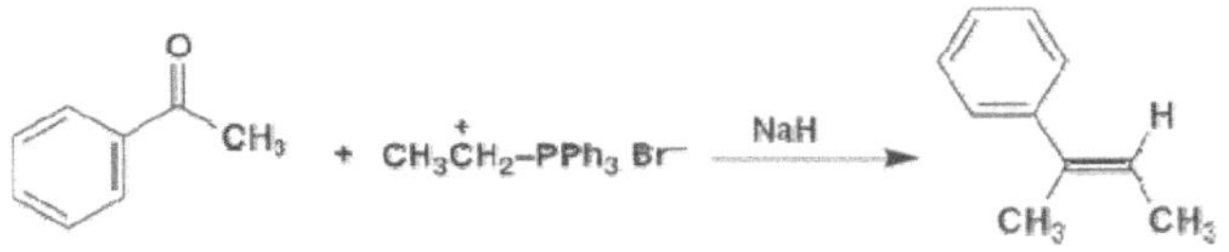 [structure: acetophenone] $+$ $CH_3\overset{+}{C}H_2\text{–}PPh_3\ Br^-$ $\xrightarrow{\text{NaH}}$ [structure: alkene product]

9. Which of the following statements is wrong?

(a) The yield of cyanohydrin is low when an aldehyde is simply added to an aqueous solution of NaCN.

(b) The yield of cyanohydrin increases when an appropriate amount of acid is added to an aqueous solution of an aldehyde and NaCN.

(c) Formation of cyanohydrin of aldehyde is accelerated by acid catalysis.

(d) Formation of cyanohydrin of aldehyde is increased by increasing the concentration of cyanide ion.

10. **Which of the following statements is wrong?**

    (a) Hydrolysis of an acetal is catalysed by acids.

    (b) Hydrolysis of an acetal is catalysed by bases.

    (c) Oximes are stabilized by conjugation between the C=N and OH groups.

    (d) Enamines are formed between secondary amines and the carbonyl group of aldehydes and ketones.

11. **Which aldehyde has the largest equilibrium constant for the formation of a cyanohydrin?**

    (a) 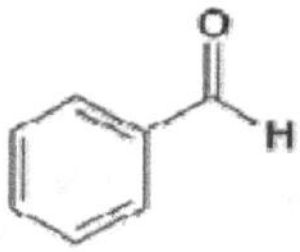   (b)

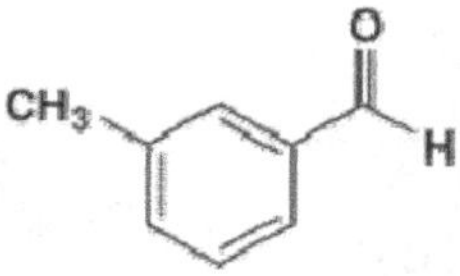

    (c) 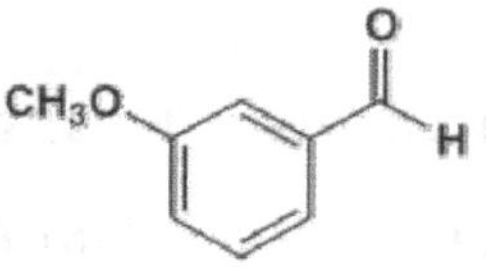   (d)

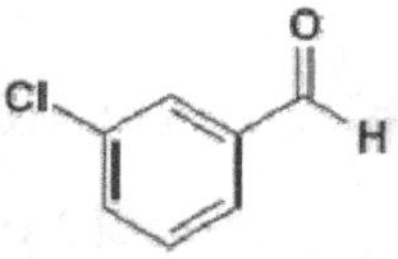

12. **Which aldehyde has the smallest equilibrium constant for the formation of a cyanohydrin?**

    (a) 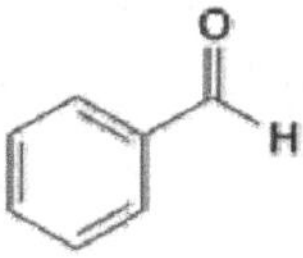   (b)

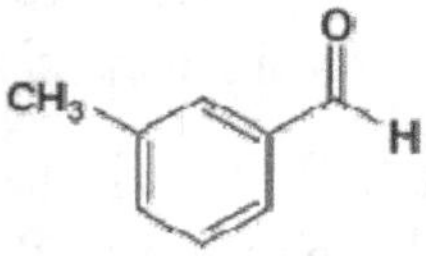

    (c) 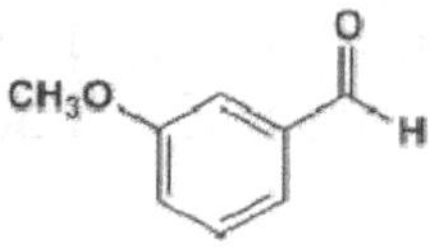   (d)

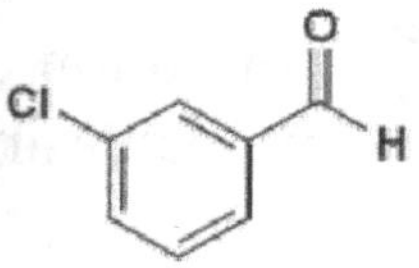

13. Which ketone has the largest equilibrium constant for hydration?

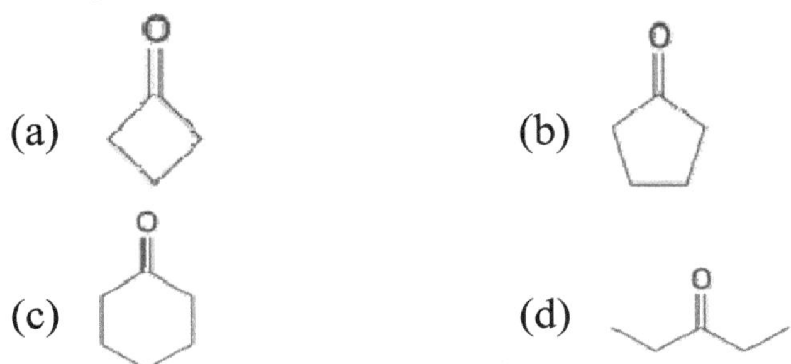

(a)

(b)

(c)

(d)

14. Which combination of an aldehyde and an alcohol most readily forms a hemiacetal with base catalysis?

(a) CH_3—CHO + CH_3CH_2OH

(b) CH_3—CHO + $(CH_3)_2CHOH$

(c) $ClCH_2$—CHO + CH_3CH_2OH

(d) $ClCH_2$—CHO + $(CH_3)_2CHOH$

15. Which of the following most readily forms a cyclic hemiacetal with acid catalysis?

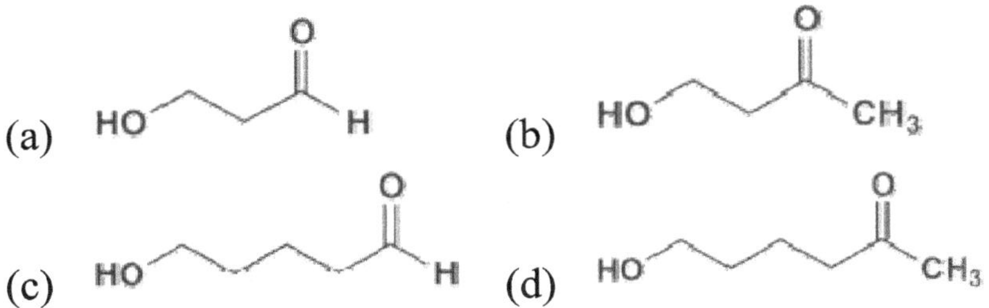

(a)

(b)

(c)

(d)

16. **Which stage of the following reaction scheme for acetal formation does not correctly show the flow of electrons?**

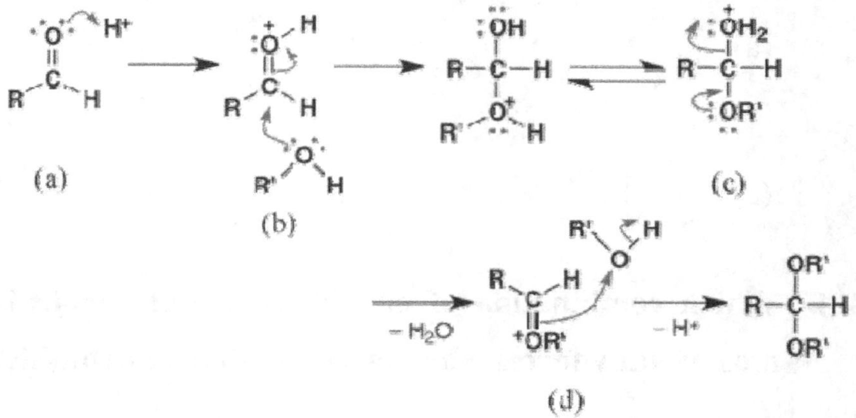

    (a)   (b)   (c)   (d)

    (a) Stage (a)   (b) Stage (b)

    (c) Stage (c)   (d) Stage (d)

17. **Which of (a)-(d) is least likely as an intermediate or a product in the following reaction?**

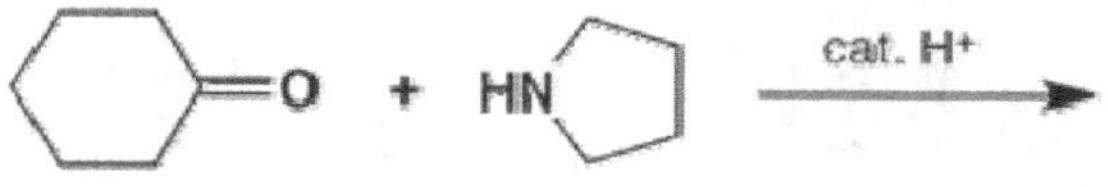

    (a) 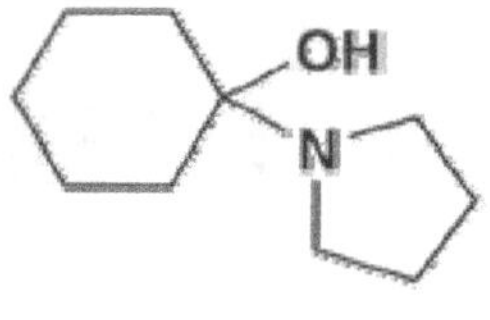   (b)

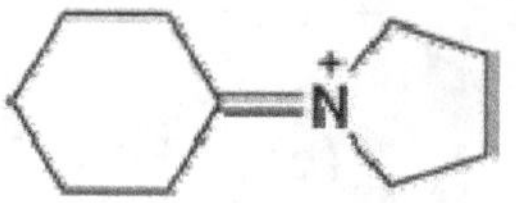

(c)   (d)

18. Which of the following reactions indicates a correct result of the main process?

(a)   cyclohexanone $=O$ + CH_3NH_2 $\xrightarrow{\text{cat. } H^+}$ cyclohexene$-NHCH_3$

(b)   cyclohexanone $=O$ + $(CH_3)_2NH$ $\xrightarrow{\text{cat. } H^+}$ $=\overset{+}{N}(CH_3)_2$

(c)   cyclohexanone $=O$ + $(CH_3)_2NH$ $\xrightarrow{\text{cat. } H^+}$ cyclohexene$-N(CH_3)_2$

(d)   cyclohexanone $=O$ + $H_2NCONHNH_2$ $\xrightarrow{\text{cat. } H^+}$ $=N-CONHNH_2$

19. Each of the following diagrams (1)-(4) indicates the flow of electrons in a possible main step in the formation of an imine from an aldehyde and an amine in aqueous solution. Which combination of two steps is the most probable in a mechanism of the overall reaction?

(1)   R'NH$_2$ (2)   R'NH$_2$ (3)   HO / NHR   (4)   H$_2$O / NHR

(a) (1) and (3)   (b) (1) and (4)

(c) (2) and (3)   (d) (2) and (4)

20. **Which of the following statements is wrong?**

(a) Hydration of an aldehyde is reversible either in acidic or basic solutions.

(b) Hydration of simple ketones is not generally favourable.

(c) Equilibrium in the hydration of aldehydes generally favours formation of the hydrate.

(d) Oxygen isotope exchange occurs when a ketone is dissolved in water labelled with ^{18}O in the presence of a catalytic amount of acid or base.

KEYS

1. (a)	**2.** (d)	**3.** (c)	**4.** (d)	**5.** (c)
6. (a)	**7.** (b)	**8.** (b)	**9.** (d)	**10.** (b)
11. (d)	**12.** (b)	**13.** (a)	**14.** (c)	**15.** (c)
16. (d)	**17.** (d)	**18.** (c)	**19.** (b)	**20.** (c)

NUCLEOPHILLIC SUBSTITUTION REACTIONS

1. With which of the following can ethyl ethanoate not undergo a substitution reaction?

   (a) aqueous NaOH

   (b) CH_3OH, H^+

   (c) aqueous NH_3

   (d) CH_3CO_2Na

2. With which of the following can ethanoic anhydride not undergo a substitution reaction?

   (a) aqueous NaOH

   (b) CH_3OH

   (c) NaCl in CH_3CO_2H

   (d) CH_3NH_2

3. Which of the following is the most reactive in dilute aqueous NaOH?

   (a) $CH_3-\overset{\overset{O}{\|}}{C}-OEt$

   (b) $CH_3-\overset{\overset{O}{\|}}{C}-Cl$

   (c) $CH_3-\overset{\overset{O}{\|}}{C}-NMe_2$

   (d) $CH_3-\overset{\overset{O}{\|}}{C}-O-\overset{\overset{O}{\|}}{C}-CH_3$

4. Which of the following is the least reactive in dilute aqueous NaOH?

   (a) $CH_3-\overset{\overset{O}{\|}}{C}-OEt$

   (b) $CH_3-\overset{\overset{O}{\|}}{C}-Cl$

   (c) $CH_3-\overset{\overset{O}{\|}}{C}-NMe_2$

   (d) $CH_3-\overset{\overset{O}{\|}}{C}-O-\overset{\overset{O}{\|}}{C}-CH_3$

5. **Which of the following is the most reactive towards methylamine?**

(a) $CH_3-\overset{O}{\overset{||}{C}}-OEt$

(b) $CH_3-\overset{O}{\overset{||}{C}}-OPh$

(c) $CH_3-\overset{O}{\overset{||}{C}}-NH_2$

(d) $CH_3-\overset{O}{\overset{||}{C}}-O-\overset{O}{\overset{||}{C}}-CH_3$

6. **Which of the following esters is the most reactive in acid-catalysed hydrolysis?**

(a) $CH_3-\overset{O}{\overset{||}{C}}-OEt$

(b) $CH_3CH_2-\overset{O}{\overset{||}{C}}-OEt$

(c) $(CH_3)_2CH-\overset{O}{\overset{||}{C}}-OEt$

(d) $(CH_3)_3C-\overset{O}{\overset{||}{C}}-OEt$

7. **Which of the following esters is the most reactive in alkaline hydrolysis?**

(a) $CH_3CH_2-\overset{O}{\overset{||}{C}}-OEt$

(b) $ClCH_2-\overset{O}{\overset{||}{C}}-OEt$

(c) $MeOCH_2-\overset{O}{\overset{||}{C}}-OEt$

(d) $H_2NCH_2-\overset{O}{\overset{||}{C}}-OEt$

8. **Which of the following esters is the least reactive in alkaline hydrolysis?**

(a) $CH_3CH_2-\overset{O}{\overset{||}{C}}-OEt$

(b) $ClCH_2-\overset{O}{\overset{||}{C}}-OEt$

(c) $MeOCH_2-\overset{O}{\overset{||}{C}}-OEt$

(d) $H_2NCH_2-\overset{O}{\overset{||}{C}}-OEt$

9. **Which of (a)-(d) indicates correctly the only organic product(s) containing the ^{18}O isotope when ethyl ethanoate is hydrolysed in an alkaline solution**

containing ^{18}O-labelled water, and neutralized with dilute hydrochloric acid?

(a) $Et^{18}OH$

(b) $Me-C(=O)-^{18}OH + Et^{18}OH$

(c) $Me-C(=O)-^{18}OH$

(d) $Me-C(=^{18}O)-OH + Me-C(=O)-^{18}OH$

10. **Which of the following statements is wrong?**

(a) For the synthesis of esters from carboxylic acids and alcohols, we can often use a base as a catalyst.

(b) For the synthesis of esters from carboxylic acids and alcohols, we can often use an additional acid as a catalyst.

(c) For the alkaline hydrolysis of carboxylic esters, we need to use an excess of base.

(d) For the hydrolysis of carboxylic esters under acidic conditions, we need only a catalytic amount of an additional acid.

11. **Which of the following is most reactive towards ethyl ethanoate?**

(a) CH_3OH

(b) CH_3O^-

(c) CH_3NH_2

(d) $CH_3CO_2^-$

12. **Which of the following is the most reactive in ethanol containing sodium ethoxide?**

(a) $Me-C(=O)-OMe$

(b) $Me-C(=O)-OPh$

(c) $Me-C(=O)-O-C(=O)-Me$

(d) $Me-C(=O)-NMe_2$

13. Which of the following is the second most reactive in ethanol containing sodium ethoxide?

(a) 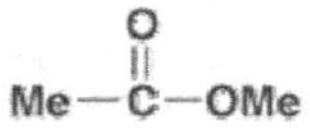 $Me-\overset{O}{\overset{\|}{C}}-OMe$

(b) $Me-\overset{O}{\overset{\|}{C}}-OPh$

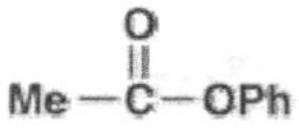

(c) $Me-\overset{O}{\overset{\|}{C}}-O-\overset{O}{\overset{\|}{C}}-Me$

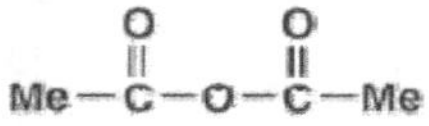

(d) 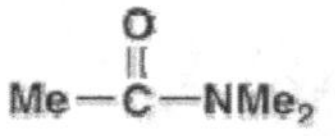 $Me-\overset{O}{\overset{\|}{C}}-NMe_2$

14. Which of the following esters is most reactive in alkaline hydrolysis?

(a)

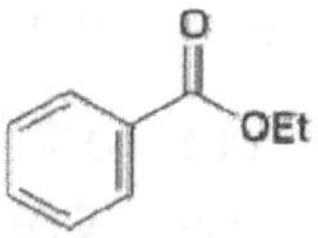

(b)

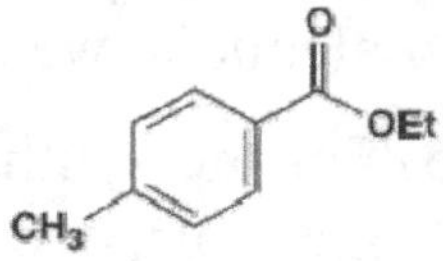

(c)

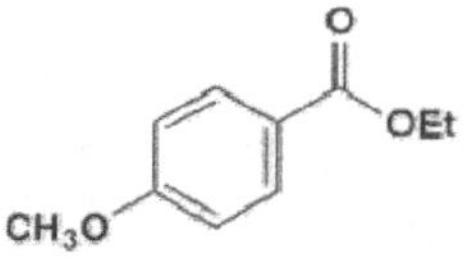

(d)

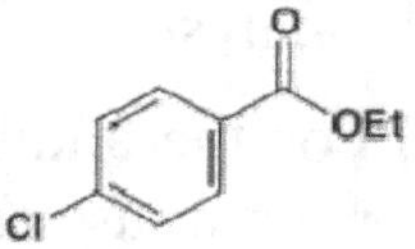

15. Which of the following esters is the least reactive in alkaline hydrolysis?

(a)

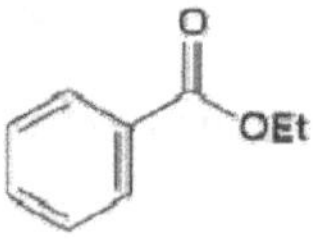

(b)

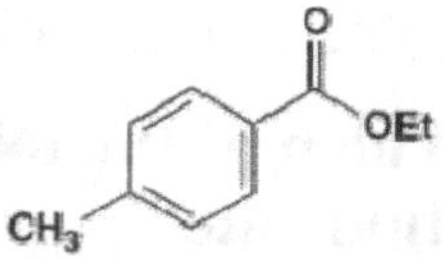

(c)

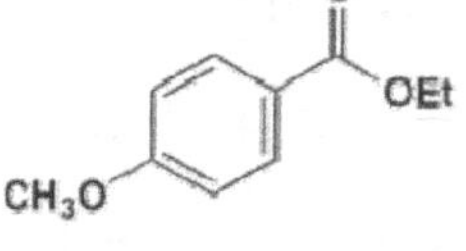

(d)

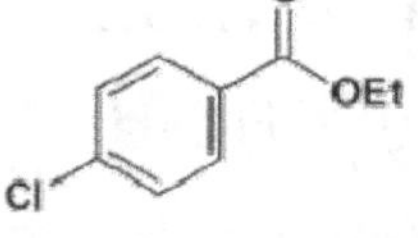

16. Which of the following within a tetrahedral intermediate leaves most readily as a nucleofuge?

(a) OH

(b) OCH_2CH_3

(c) $OCOCH_3$

(d) NH_2

17. Which of the following tetrahedral intermediates (formed in nucleophilic substitution reactions of carboxylic acid derivatives) gives a compound not obtained from others as a main product?

(a) 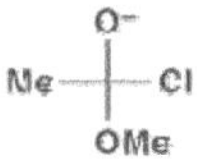 Me—Cl, OMe, O⁻

(b) 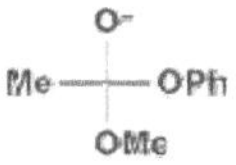 Me—OPh, OMe, O⁻

(c) 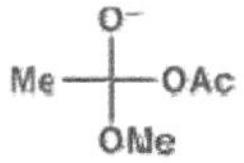 Me—OAc, OMe, O⁻

(d) 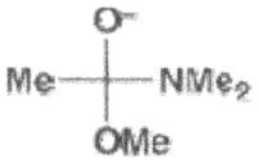 Me—NMe₂, OMe, O⁻

18. Which stage of the following reaction scheme for the acid-catalysed hydrolysis of an ester does not correctly show the flow of electrons?

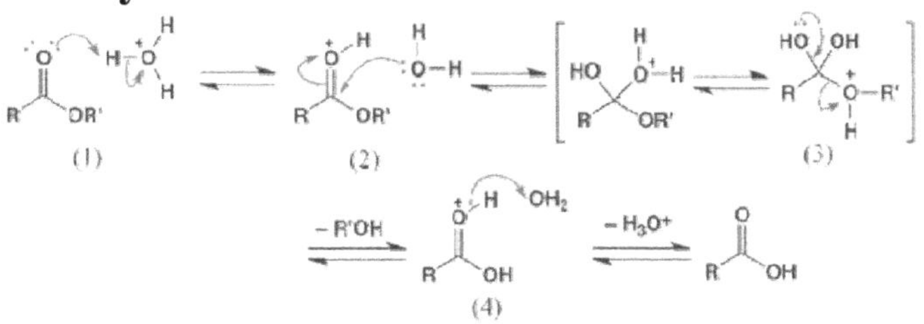

(1)   (2)   (3)   (4)

(a) Stage (1)   (b) Stage (2)
(c) Stage (3)   (d) Stage (4)

19. Which of the following statements regarding acid-catalysed hydrolysis of carboxylic esters is wrong?

(a) Protonation of the carbonyl oxygen enhances its electrophilicity.

(b) Hydroxide ion adds to the protonated carbonyl group.

(c) Protonation of the alkoxy oxygen of the tetrahedral intermediate facilitates elimination of the alcohol.

(d) The initially formed protonated carboxylic acid is deprotonated to give the final product.

20. **Which of the following statements regarding the hydrolysis of labelled ethyl ethanoates, 1 or 2, in normal water is wrong?**

$$\underset{\text{Me}}{\overset{^{18}O}{\|}}\text{OEt} \qquad \underset{\text{Me}}{\overset{O}{\|}}\,^{18}\text{OEt}$$

1   2

(a) During the alkaline hydrolysis of 1, unlabelled ethyl ethanoate can be recovered.

(b) During the acid hydrolysis of 1, unlabelled ethyl ethanoate can be recovered.

(c) The product ethanoic acid contains the ^{18}O isotope in acid hydrolysis of both 1 and 2.

(d) The product ethanol contains the ^{18}O isotope in alkaline hydrolysis of 2.

<u>**KEYS**</u>

1. (d)	**2.** (c)	**3.** (b)	**4.** (c)	**5.** (d)
6. (a)	**7.** (b)	**8.** (a)	**9.** (d)	**10.** (a)
11. (b)	**12.** (c)	**13.** (b)	**14.** (d)	**15.** (c)
16. (c)	**17.** (d)	**18.** (d)	**19.** (b)	**20.** (c)

CHAPTER 10

REACTIONS OF CARBONYL COMPOUND

1. Which is unreactive in hydride reduction with $NaBH_4$?

   (a) 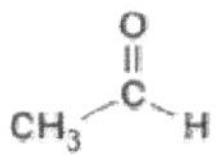 CH_3—$\overset{\overset{O}{\|}}{C}$—$H$

   (b) CH_3—$\overset{\overset{O}{\|}}{C}$—$CH_3$

   (c) CH_3—$\overset{\overset{O}{\|}}{C}$—$OCH_3$

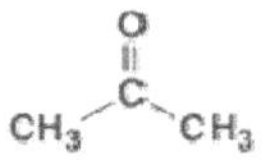

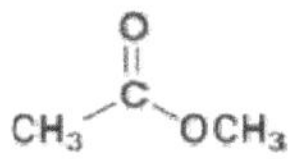

   (d) 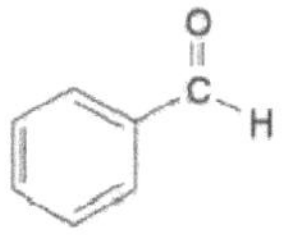 (benzaldehyde)

2. Which is the reduction product of ethyl 3-oxobutanoate with $NaBH_4$ in methanol?

   (a)

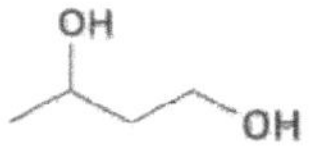

   (b)

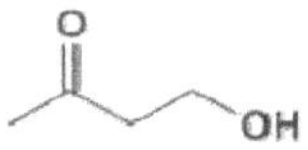

   (c)

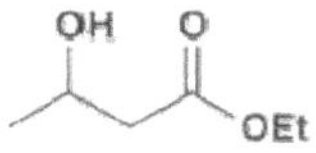

   (d)

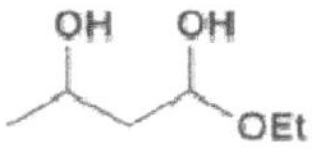

3. Which is the main reduction product of ethyl 3-oxobutanoate with $LiAlH_4$ in ether followed by an aqueous workup?

   (a)

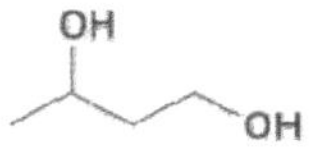

   (b)

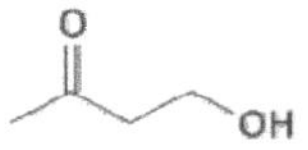

   (c)

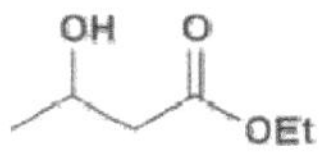

   (d)

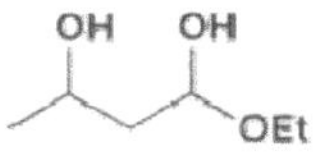

4. **Which is the reduction product of *N*-methyl propanamide with LiAlH₄ in ether followed by an aqueous workup?**

   (a) $CH_3CH_2\overset{\overset{\displaystyle OH}{|}}{C}HNHCH_3$

   (b) $CH_3CH_2CH_2NHCH_3$

   (c) $CH_3CH_2CH_2OH$

   (d) $CH_3CH_2CH_2NH_2$

5. **Which pair of reactants for a Grignard reaction does not give 2-phenylbutan-2-ol after an aqueous workup?**

   (a) 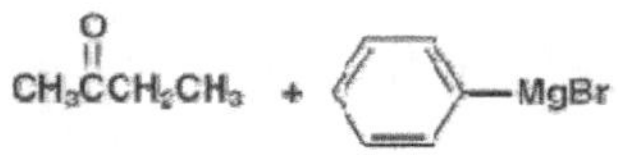 $CH_3\overset{\overset{\displaystyle O}{\|}}{C}CH_2CH_3$ + Ph–MgBr

   (b) 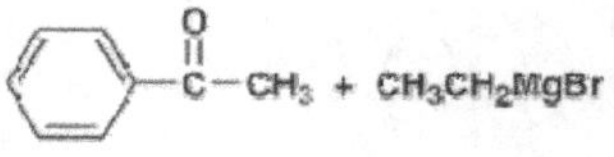 Ph–$\overset{\overset{\displaystyle O}{\|}}{C}$–CH₃ + CH₃CH₂MgBr

   (c) Ph–$\overset{\overset{\displaystyle O}{\|}}{C}$–CH₂CH₃ + CH₃MgBr

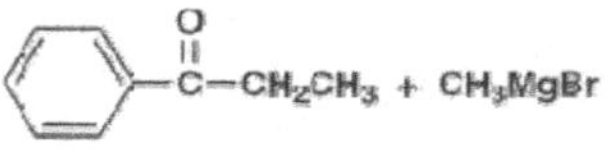

   (d) 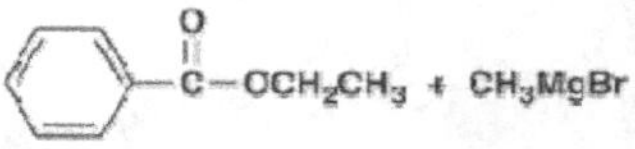 Ph–$\overset{\overset{\displaystyle O}{\|}}{C}$–OCH₂CH₃ + CH₃MgBr

6. **Which pair of reactants for a Grignard reaction does not give triphenylmethanol after an aqueous workup?**

   (a) Ph–$\overset{\overset{\displaystyle O}{\|}}{C}$–H + 2 PhMgBr

   (b) Ph–$\overset{\overset{\displaystyle O}{\|}}{C}$–OMe + 2 PhMgBr

   (c) Ph–$\overset{\overset{\displaystyle O}{\|}}{C}$–Ph + PhMgBr

   (d) MeO–$\overset{\overset{\displaystyle O}{\|}}{C}$–OMe + 3 PhMgBr

7. **Which of the following reactions does not give a secondary alcohol?**

   (a) $CH_3\overset{\overset{\displaystyle O}{\|}}{C}CH_2CH_3$ + NaBH₄ $\xrightarrow{\text{MeOH}}$

   (b) Ph–$\overset{\overset{\displaystyle O}{\|}}{C}$–H + CH₃MgBr $\xrightarrow[\text{2) H}_3\text{O}^+]{\text{1) Et}_2\text{O}}$

   (c) $CH_3\overset{\overset{\displaystyle O}{\|}}{C}CH_2CH_3$ + LiAlH₄ $\xrightarrow[\text{2) H}_3\text{O}^+]{\text{1) Et}_2\text{O}}$

   (d) $CH_3\overset{\overset{\displaystyle O}{\|}}{C}OC_2H_5$ + CH₃MgBr $\xrightarrow[\text{2) H}_3\text{O}^+]{\text{1) Et}_2\text{O}}$

8. Which of the following reactions does not give a primary alcohol?

(a) $(CH_3)_2CHCH{=}O$ + $NaBH_4$ $\xrightarrow{MeOH}$

(b) $CH_3COCH_2CH_3$ + $NaBH_4$ $\xrightarrow{MeOH}$

(c) $CH_3COCH_2CH_3$ + $LiAlH_4$ $\xrightarrow[2)\ H_3O^+]{1)\ Et_2O}$

(d) $H_2C{=}O$ + C_2H_5MgBr $\xrightarrow[2)\ H_3O^+]{1)\ Et_2O}$

9. Which of the following reactions does not give the product indicated?

(a) $Ph{-}C(=O){-}H$ + $PhMgBr$ $\xrightarrow[2)\ H_3O^+]{1)\ Et_2O}$ Ph_2CHOH

(b) $Ph{-}C(=O){-}OEt$ + $2\ CH_3MgI$ $\xrightarrow[2)\ H_3O^+]{1)\ Et_2O}$ $PhC(OH)(CH_3)_2$

(c) 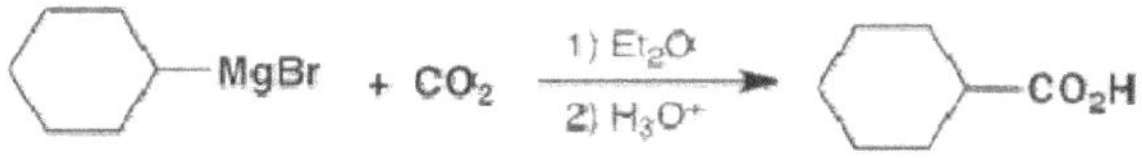 ⬡—$MgBr$ + CO_2 $\xrightarrow[2)\ H_3O^+]{1)\ Et_2O}$ ⬡—CO_2H

(d) 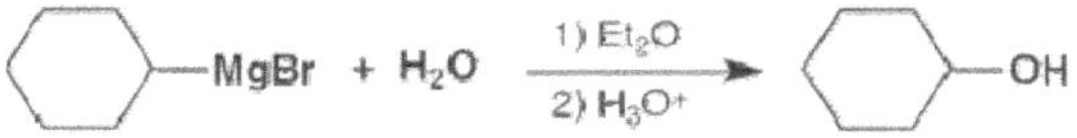 ⬡—$MgBr$ + H_2O $\xrightarrow[2)\ H_3O^+]{1)\ Et_2O}$ ⬡—OH

10. Which of the following reactions gives a product different from that of the other reactions?

(a) $CH_3CH_2CH{=}O$ + CH_3CH_2MgBr $\xrightarrow[2)\ H_3O^+]{1)\ Et_2O}$

(b) $CH_3CH_2COCH_2CH_3$ + $NaBH_4$ $\xrightarrow{MeOH}$

(c) $CH_3CH_2COCH_3$ + CH_3CH_2MgBr $\xrightarrow[2)\ H_3O^+]{1)\ Et_2O}$

(d) $HCOCH_2CH_3$ + CH_3CH_2MgBr $\xrightarrow[2)\ H_3O^+]{1)\ Et_2O}$

11. Which is the main reduction product of the following compound with NaBH₄ in methanol?

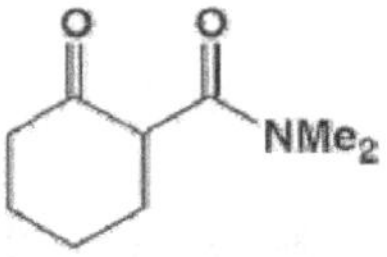

(a)

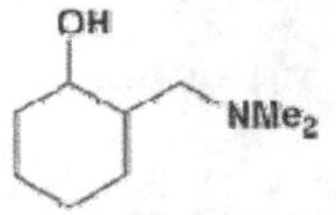

(b)

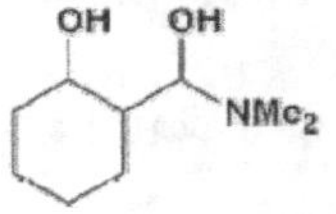

(c)

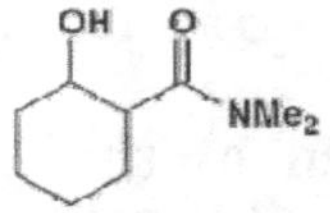

(d)

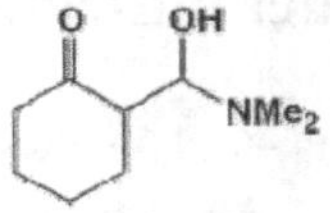

12. Which is the main reduction product of the following compound with LiAlH₄ in ether followed by an aqueous workup?

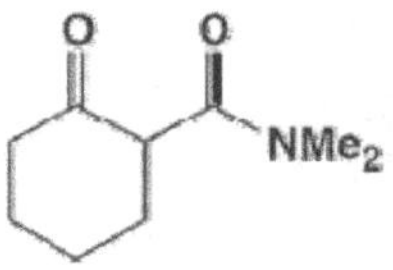

(a)

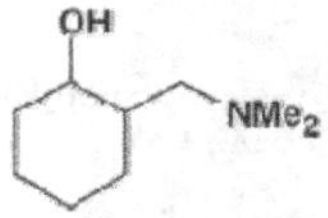

(b)

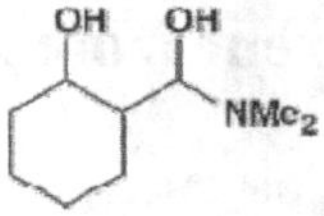

(c)

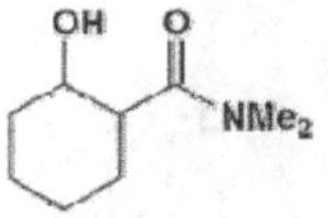

(d)

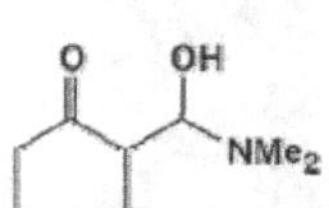

13. **Which is the main reduction product of the following compound with NaBH$_4$ in methanol?**

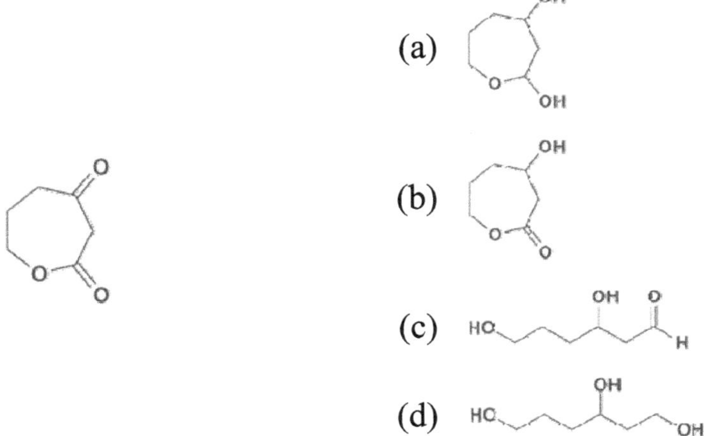

14. **Which is the reduction product of the following compound with LiAlH$_4$ in ether followed by aqueous workup?**

15. **Which of the following incorrectly represents the flow of electrons in the hydride reduction of an aldehyde with NaBH$_4$?**

(a) 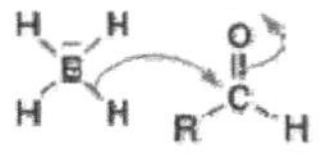   (b)

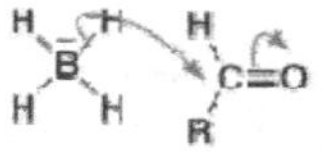

(c) 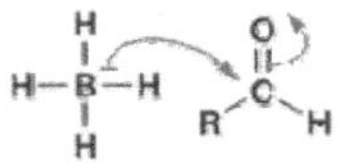   (d)

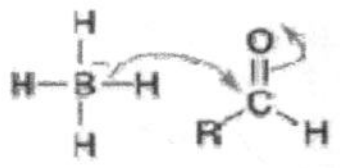

16. **Which of the following carbonyl compounds does not give 2-methylpentan-2-ol upon reaction with methyl magnesium iodide in ether followed by a workup with aqueous ammonium chloride?**

(a) pentan-2-one
(b) methyl propyl ketone
(c) ethyl butanoate
(d) ethyl 2-methyl butanoate

17. **Which of the following reactions does not give a primary alcohol?**

(a) $PhCH_2CO_2Et$ + $NaBH_4$ $\xrightarrow{MeOH}$

(b) $PhCO_2Et$ + $LiAlH_4$ $\xrightarrow[\text{2) } H_2O^+]{\text{1) } Et_2O}$

(c)  $PhMgBr$ + (epoxide) $\xrightarrow[\text{2) } H_3O^+]{\text{1) } Et_2O}$

(d) $PhCH_2CHO$ + $NaBH_4$ $\xrightarrow{MeOH}$

18. **Which compound does not give a tertiary alcohol upon reaction with ethylmagnesium bromide in ether followed by a workup with aqueous ammonium chloride?**

(a) 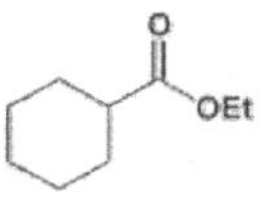   (b) 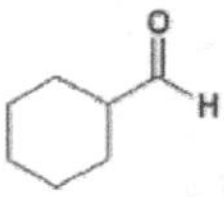   (c) 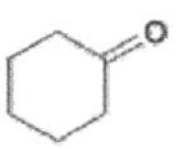   (d)

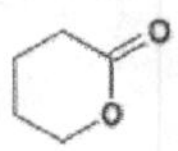

19. **Which of the following reactions does not give the product indicated?**

(a) $MeOCH_2COEt$ + $LiAlH_4$ $\xrightarrow[\text{2) } H_3O^+]{\text{1) } Et_2O}$ $MeOCH_2CH_2OH$

(b) $MeOCH_2COEt$ + CH_3MgBr $\xrightarrow[\text{2) } H_3O^+]{\text{1) } Et_2O}$ $MeOCH_2C(CH_3)_2OH$

(c) $MeOCH_2CNMe_2$ + CH_3MgBr $\xrightarrow[\text{2) } H_3O^+]{\text{1) } Et_2O}$ $MeOCH_2CCH_3$

(d) $MeOCH_2CNMe_2$ + $LiAlH_4$ $\xrightarrow[\text{2) } H_3O^+]{\text{1) } Et_2O}$ $MeOCH_2CH$

20. **Which of the following reactions does not give the product indicated?**

(a) $PhMgBr$ + CH_3CHO $\xrightarrow[\text{2) } H_3O^+]{\text{1) } Et_2O}$ $Ph-\underset{H}{\overset{CH_3}{C}}-OH$

(b) $PhMgBr$ + $CH_3COOC_2H_5$ $\xrightarrow[\text{2) } H_3O^+]{\text{1) } Et_2O}$ $Ph-\underset{C_2H_5}{\overset{CH_3}{C}}-OH$

(c) $PhMgBr$ + CO_2 $\xrightarrow[\text{2) } H_3O^+]{\text{1) } Et_2O}$ $PhCO_2H$

(d) $PhMgBr$ + D_2O $\xrightarrow[\text{2) } H_3O^+]{\text{1) } Et_2O}$ C_6H_5-D

KEYS

1. (c)	**2.** (c)	**3.** (a)	**4.** (b)	**5.** (d)
6. (a)	**7.** (d)	**8.** (b)	**9.** (d)	**10.** (c)
11. (c)	**12.** (a)	**13.** (b)	**14.** (d)	**15.** (c)
16. (d)	**17.** (a)	**18.** (b)	**19.** (d)	**20.** (b)

CHAPTER 11

HALOALKANES

1. **Which of the following statements regarding the S_N2 mechanism is wrong?**

    (a) S_N2 reactions are bimolecular.
    (b) S_N2 reactions are usually second order.
    (c) The S_N2 mechanism occurs in one step.
    (d) S_N2 reactions usually occur in two steps.

2. **Which of the following statements regarding the S_N1 mechanism is wrong?**

    (a) S_N1 reactions are unimolecular.
    (b) S_N1 reactions are first order.
    (c) The S_N1 mechanism involves a single step.
    (d) S_N1 reactions usually occur in two steps.

3. **Which is the most reactive compound by the S_N2 mechanism?**

    (a) $CH_3CH_2CH_2CH_2Br$ (b) $(CH_3)_2CHCH_2Br$

    (c) $CH_3CH_2\overset{\overset{\displaystyle Br}{|}}{C}HCH_3$ (d) $CH_3CH_2\overset{\overset{\displaystyle Br}{|}}{C}(CH_3)_2$

4. **Which is the least reactive compound by the S_N2 mechanism?**

    (a) $CH_3CH_2CH_2CH_2Br$ (b) $(CH_3)_2CHCH_2Br$

(c) 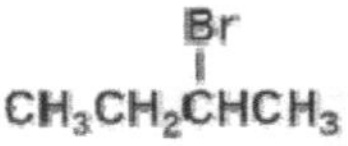 $CH_3CH_2\overset{\overset{\displaystyle Br}{|}}{C}HCH_3$

(d) 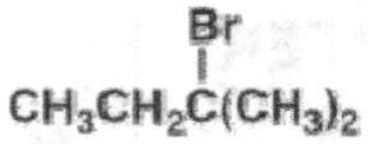 $CH_3CH_2\overset{\overset{\displaystyle Br}{|}}{C}(CH_3)_2$

5. **Which is the most reactive compound by the S_N1 mechanism?**

(a) $CH_2=CHCH_2Br$

(b) 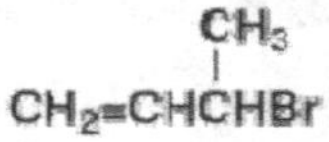 $CH_2=CH\overset{\overset{\displaystyle CH_3}{|}}{C}HBr$

(c) $CH_3CH_2CH_2Br$

(d) 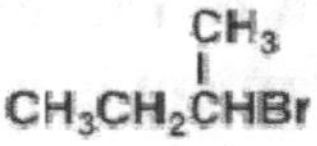 $CH_3CH_2\overset{\overset{\displaystyle CH_3}{|}}{C}HBr$

6. **Which is the least reactive compound by the S_N1 mechanism?**

(a) $CH_2=CHCH_2Br$

(b) 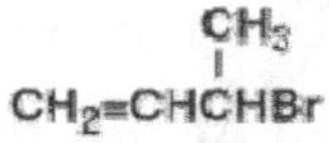 $CH_2=CH\overset{\overset{\displaystyle CH_3}{|}}{C}HBr$

(c) $CH_3CH_2CH_2Br$

(d) 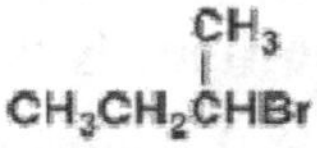 $CH_3CH_2\overset{\overset{\displaystyle CH_3}{|}}{C}HBr$

7. **Which is the most reactive compound by the S_N1 mechanism?**

(a) 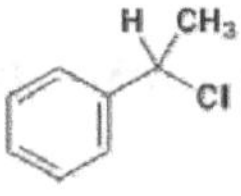 (b) 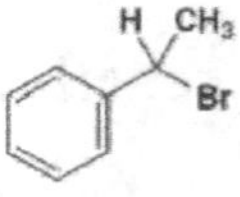 (c) 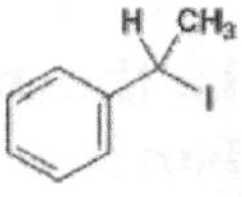 (d)

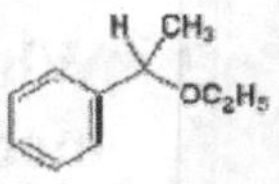

8. **Which is the least reactive compound by the S_N1 mechanism?**

(a) 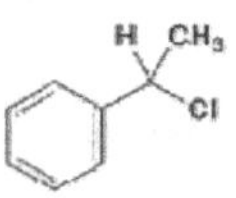 (b) 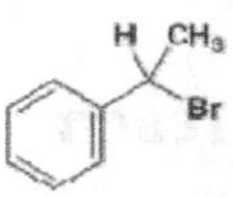 (c) 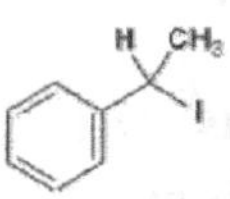 (d)

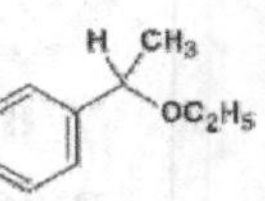

9. **Which is the main product of the following reaction?**

+ $NaSC_6H_5$ → propanone

(a) SC_6H_5 (b) SC_6H_5 (c) SC_6H_5 (d)

10. **Which is the main product of the following reaction?**

I ~~~ Cl + NaCN → MeOH

(a) NC ~~~ I (b) NC ~~~ Cl

(c) MeO ~~~ Cl (d) ~~~ Cl

11. **Which of the following statements is wrong?**

(a) S_N1 reactions undergo partial inversion of configuration.
(b) S_N2 reactions undergo partial inversion of configuration.
(c) The rate constant of an S_N1 reaction depends on the solvent.
(d) The rate constant of an S_N2 reaction depends on the solvent.

12. **Which of the following statements is wrong?**

(a) S_N1 reactions proceed via carbenium ion inter mediates.
(b) The S_N2 mechanism does not involve an inter mediate.
(c) The rate constant of an S_N1 reaction depends on the nucleofuge.
(d) The rate constant of an S_N2 reaction does not depend on the nucleofuge.

13. **Which of the following statements regarding nucleophilicity is wrong?**

    (a) Ethoxide ion is more nucleophilic than *t*-butoxide in spite of its lower basicity.

    (b) Ethoxide ion is more nucleophilic than *t*-butoxide due to the lower steric hindrance.

    (c) Chloride ion is more nucleophilic than iodide ion because of its higher basicity.

    (d) Bromide ion is more nucleophilic than fluoride in spite of its lower basicity.

14. **Which of the following is most reactive as a nucleophile?**

    (a) PhO^-

    (b) PhS^-

    (c) $PhCH_2O^-$

    (d) $PhCH_2NH_2$

15. **Which of the following is least basic?**

    (a) PhO^-

    (b) PhS^-

    (c) $PhCH_2O^-$

    (d) $PhCH_2NH_2$

16. **Which is the most reactive compound by the S_N1 mechanism?**

    (a) 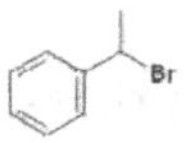 1-phenylethyl bromide   (b) 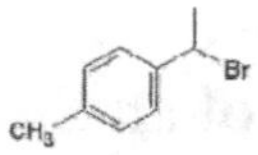 1-(4-methylphenyl)ethyl bromide   (c) 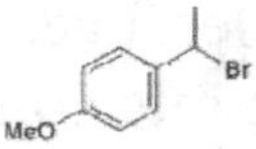 1-(4-methoxyphenyl)ethyl bromide   (d) 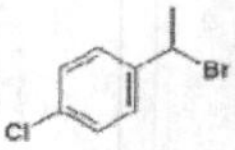 1-(4-chlorophenyl)ethyl bromide

17. **Which is the least reactive compound by the S_N1 mechanism?**

(a) 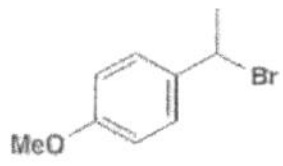   (b) 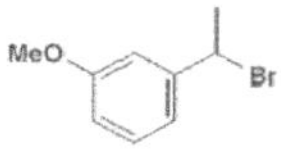   (c) 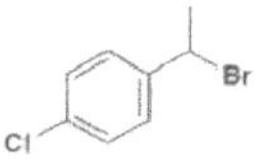   (d)

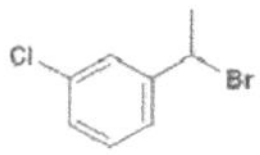

18. **Which is the most reactive compound by the S_N2 mechanism?**

(a) 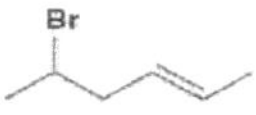   (b) 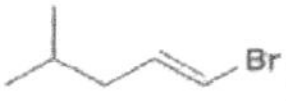   (c)    (d)

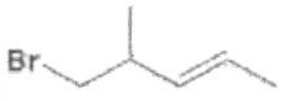

19. **Which is the least reactive compound by the S_N2 mechanism?**

(a) 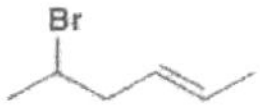   (b)

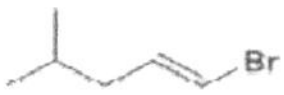

(c) 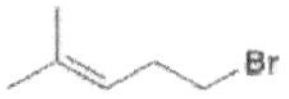   (d)

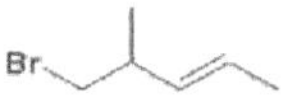

20. **Which compound is the main product of the following reaction?**

(a)    (b)

(c)    (d)

21. **Which of the following statements regarding the E1 mechanism is wrong?**

(a) Reactions by the E1 mechanism are unimolecular in the rate-determining step.

(b) Reactions by the E1 mechanism are generally first order.

(c) Reactions by the E1 mechanism usually occur in one step.

(d) Reactions by the E1 mechanism are multi-step reactions.

22. **Which of the following statements regarding the E2 mechanism is wrong?**
    (a) Reactions by the E2 mechanism are always bi molecular.
    (b) Reactions by the E2 mechanism are generally second order.
    (c) Reactions by the E2 mechanism usually occur in one step.
    (d) Reactions by the E2 mechanism usually occur in two steps.

23. **Which of the following reacts by the E1 mechanism in ethanol most readily?**
    (a) $CH_3CH_2CH_2CH_2Br$
    (b) $(CH_3)_2CHCH_2Br$
    (c) $(CH_3)_3CBr$
    (d) $CH_3CH_2CHBrCH_3$

24. **Which is the main product of the following reaction?**

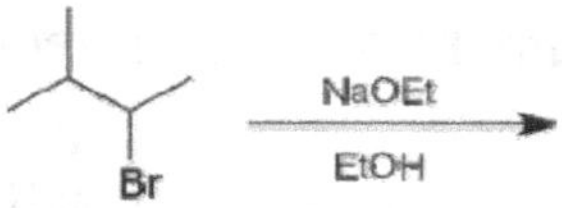

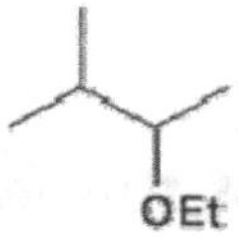

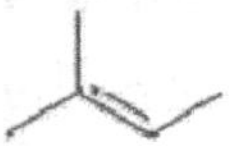

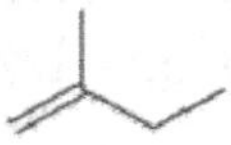

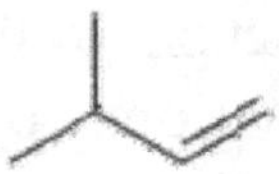

25. **Which is the main product of the following reaction?**

(a) 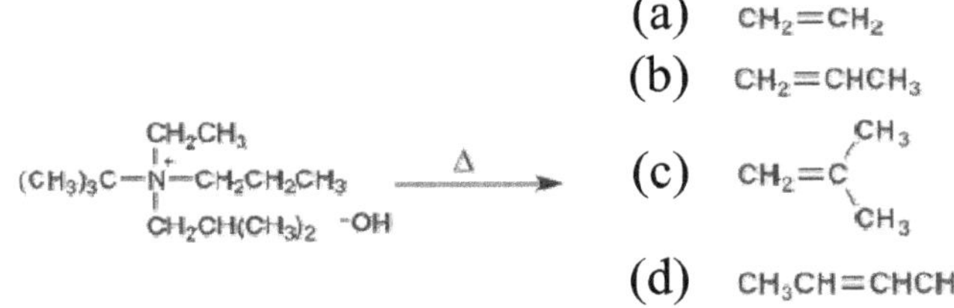

(b)

(c)

(d)

26. **Elimination reactions of compounds 1 and 2 with sodium ethoxide give two alkenes, A and B, but with different selectivities. Which of the following statements best indicates the most probable outcome?**

(a) A both from 1and 2   (b) B both from 1 and 2

(c) A from 1 and B from 2 (d) B from 1 and A from 2

27. **Which is the main alkene product in the following reaction?**

(a) $CH_2=CH_2$

(b) $CH_2=CHCH_3$

(c) $CH_2=C\begin{smallmatrix}CH_3\\CH_3\end{smallmatrix}$

(d) $CH_3CH=CHCH_3$

28. **Elimination reactions of *cis*- and *trans*-1-bromo-2-methylcyclohexanes with NaOEt in EtOH can give the same or different main product, 1-methylcyclohexene (1) or 3-methylcyclohexene (2). Which of (a)-(d) indicates the main products?**

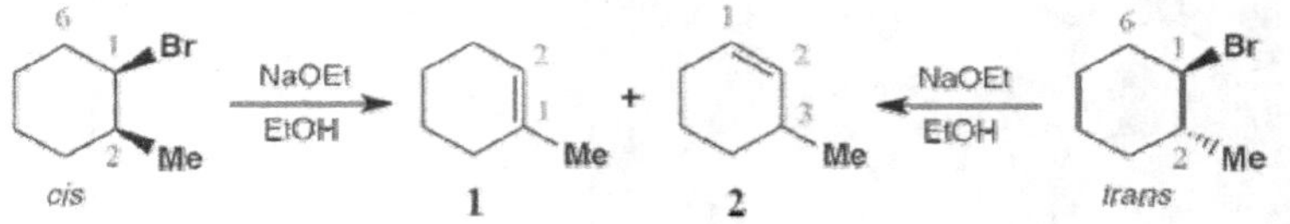

(a) 1 both from the *cis* and *trans* substrates

(b) 2 both from the *cis* and *trans* substrates

(c) 1 from the *cis* and 2 from the *trans* substrate

(d) 2 from the *cis* and 1 from the *trans* substrate

29. **Which is the main product of the following reaction?**

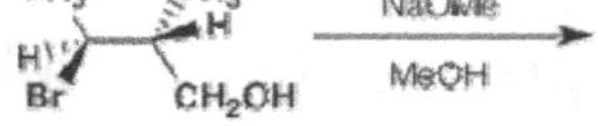

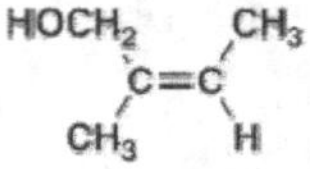

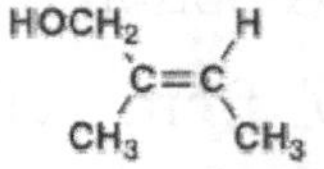

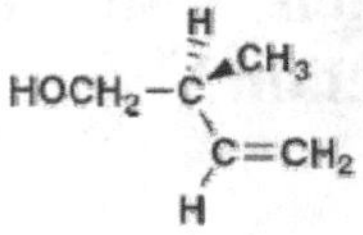

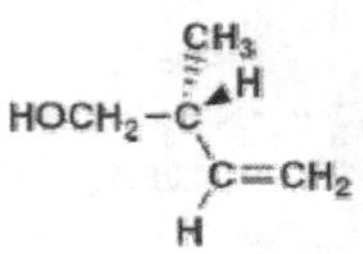

30. Which is the main product of the following reaction?

(a)
$$HOCH_2, CH_3 \\ C=C \\ CH_3, H$$

(b)
$$HOCH_2, H \\ C=C \\ CH_3, CH_3$$

(c)
$$HOCH_2-C(H)(CH_3)-C(H)=CH_2$$

(d)
$$HOCH_2-C(CH_3)(H)-C(H)=CH_2$$

Reactant:

$$\begin{array}{c} CH_3 \\ H-\!\!\!\!-Br \\ CH_3-\!\!\!\!-H \\ CH_2OH \end{array} \xrightarrow[\text{MeOH}]{\text{NaOMe}}$$

31. Which of the following statements regarding mechanisms of elimination reaction is wrong?

(a) The E1 mechanism does not require a base.

(b) The E2 mechanism generally occurs under highly basic conditions.

(c) The E2 mechanism is stereospecific.

(d) The E1 mechanism is usually unimolecular in the rate-determining step but leads to a   second order rate law.

32. Which of the following statements regarding regioselectivity of elimination reactions is wrong?

(a) More substituted, more stable alkenes are generally formed preferentially by both E1 and E2 mechanisms.

(b) Substrates with a poorer nucleofuge tend to give the less substituted alkenes.

(c) Sterically hindered, bulky bases tend to give the less substituted alkenes.

(d) Reactions by the E1 mechanism are generally less regioselective than those by the E2 mechanism.

33. **Which compound is most reactive in solvolysis (E1 and S_N1) in ethanoic acid?**

(a) 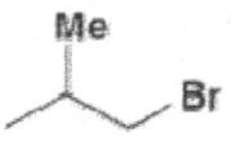 (b) 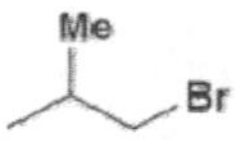 (c) 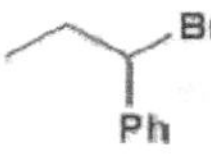 (d)

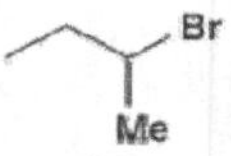

34. **Which is the main product of the following reaction?**

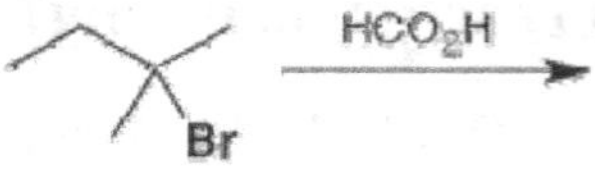

(a)

(b)

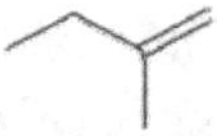

(c)

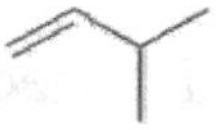

(d)

35. **Which is the main product of the following reaction?**

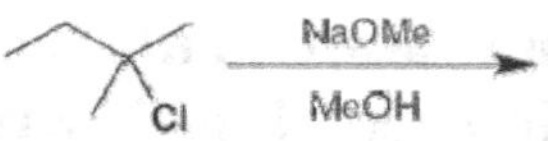

(a)

(b)

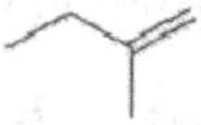

(c)

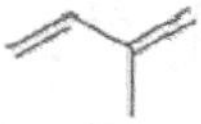

(d)

36. Which is the main product of the following reaction?

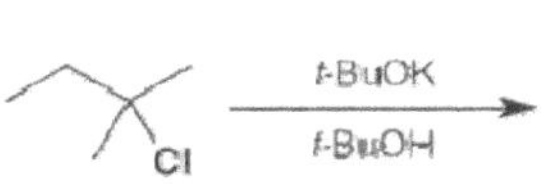

(reaction: isobutyl-type chloride with *t*-BuOK / *t*-BuOH)

(a) 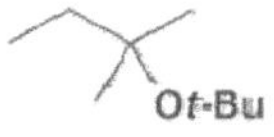 structure with O*t*-Bu

(b)  alkene structure

(c) 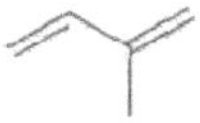 alkene structure

(d) 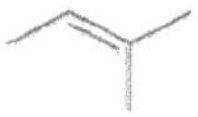 alkene structure

37. Which of the compounds bearing different nucleofuges Y in (a)-(d) gives the terminal alkene most selectively in the following reaction?

(reaction: substrate with Y, NaOEt / EtOH → internal alkene + terminal alkene)

(a) Y = F　　　　　　　　　　(b) Y = Cl

(c) Y = Br　　　　　　　　　(d) Y = OSO_2CF_3

38. Which is the main product of the following reaction?

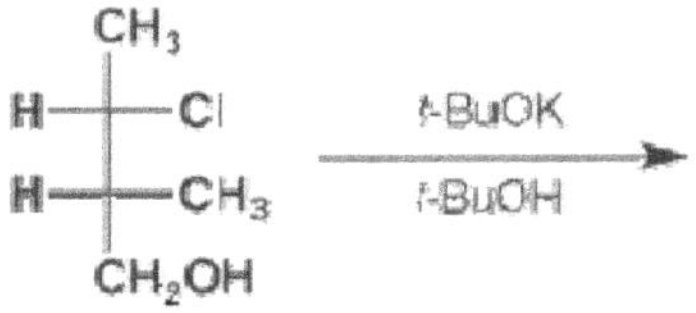

(Fischer projection: CH_3 top; H—Cl; H—CH_3; CH_2OH bottom; with *t*-BuOK / *t*-BuOH)

(a) 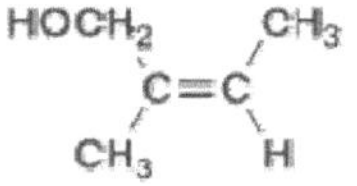
$$HOCH_2 \quad CH_3$$
$$\diagdown C = C \diagup$$
$$CH_3 \qquad H$$

(b) 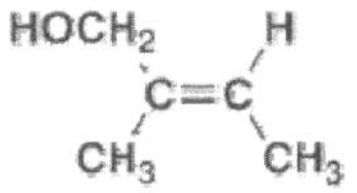
$$HOCH_2 \quad H$$
$$\diagdown C = C \diagup$$
$$CH_3 \qquad CH_3$$

(c) 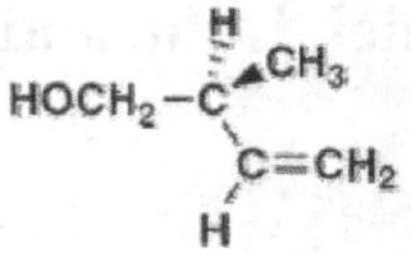 [structure: HOCH₂–C(H)(CH₃)–C=CH₂ with H]

(d) 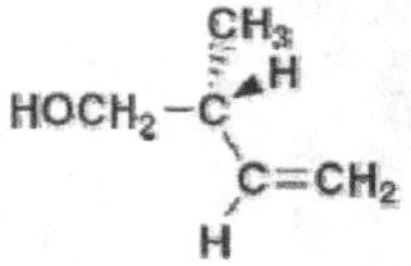 [structure: HOCH₂–C(CH₃)(H)–C=CH₂ with H]

39. **Reactions of dideuterated bromocyclohexanes, 1 and 2, by the E2 mechanism give monodeuterated cyclohexene A and dideuterated cyclohexene B. Which of the statements ((a)-((d) best indicates the most probable selectivities of two reactions.**

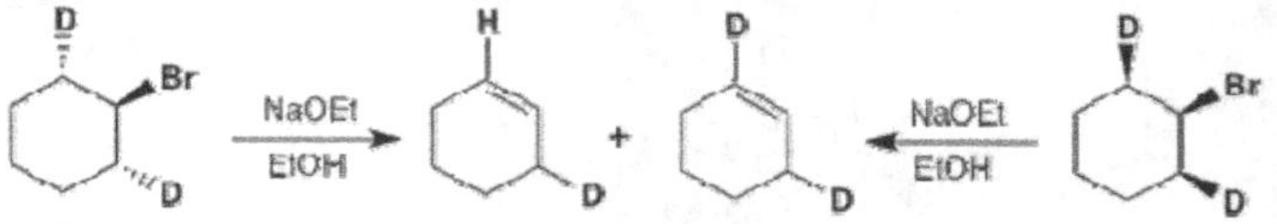

[Reaction scheme: compound 1 with NaOEt/EtOH gives A + B; compound 2 with NaOEt/EtOH]

(a) A both from 1and 2   (b) B both from 1 and 2
(c) A from 1 and B from 2   (d) B from 1 and A from 2

40. **Which is the final main product of the following reaction of *trans*-1,2-dibromocyclohexane?**

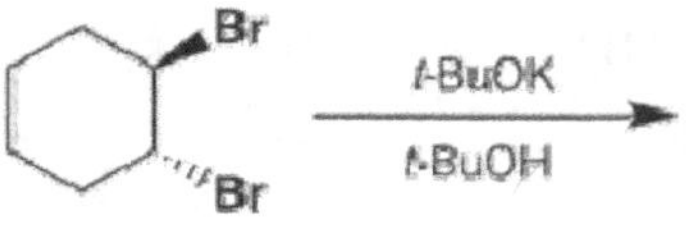

[Reaction: trans-1,2-dibromocyclohexane with t-BuOK / t-BuOH]

(a) 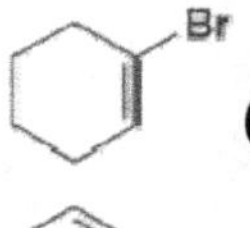 [1-bromocyclohexene] (b) 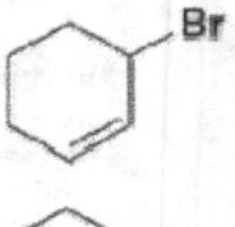 [3-bromocyclohexene]

(c)  [1,3-cyclohexadiene] (d)  [cyclohexene structure]

<u>KEYS</u>

1. (d)	**2.** (c)	**3.** (a)	**4.** (d)	**5.** (b)
6. (c)	**7.** (c)	**8.** (d)	**9.** (a)	**10.** (b)
11. (b)	**12.** (d)	**13.** (c)	**14.** (b)	**15.** (b)
16. (c)	**17.** (d)	**18.** (c)	**19.** (b)	**20.** (a)
21. (c)	**22.** (d)	**23.** (c)	**24.** (b)	**25.** (d)
26. (d)	**27.** (a)	**28.** (c)	**29.** (a)	**30.** (b)
31. (a)	**32.** (d)	**33.** (c)	**34.** (d)	**35.** (d)
36. (b)	**37.** (a)	**38.** (c)	**39.** (c)	**40.** (c)

ALCOHOLS, ETHERS, THIOLS, SULFIDES AND AMINES

1. Which compound is the most reactive in hydrochloric acid?

   (a) CH_2OH (cyclohexylmethanol)

   (b) OH, CH_3 (1-methylcyclohexanol)

   (c) CH_3, OH (4-methylcyclohexanol)

   (d) OH, CH_3 (2-methylcyclohexanol)

2. Which compound is the least reactive in hydrochloric acid?

   (a) CH_2OH (cyclohexylmethanol)

   (b) OH, CH_3 (1-methylcyclohexanol)

   (c) CH_3, OH (4-methylcyclohexanol)

   (d) OH, CH_3 (2-methylcyclohexanol)

3. Which compound is the most reactive in acid-catalysed dehydration?

   (a) $\underset{\text{CH}_3}{\text{CH}_3\text{CHCH}_2\text{CH}_2\text{OH}}$

   (b) $\underset{\text{OH}}{\overset{\text{CH}_3}{\text{CH}_3\text{CHCHCH}_3}}$

   (c) $\underset{\text{OH}}{\overset{\text{CH}_3}{\text{CH}_3\text{CCH}_2\text{CH}_3}}$

   (d) $\underset{\text{CH}_3}{\text{HOCH}_2\text{CHCH}_2\text{CH}_3}$

4. Which are the main products when methyl phenyl ether (anisole) reacts with concentrated hydriodic acid?

(a) 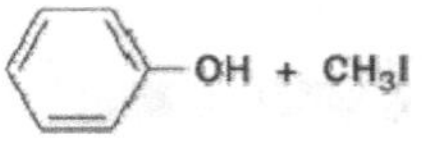 ⬡—OH + CH$_3$I (b) 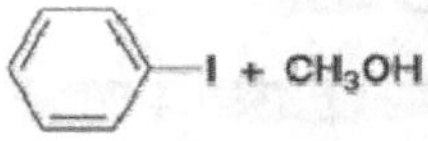 ⬡—I + CH$_3$OH

(c) 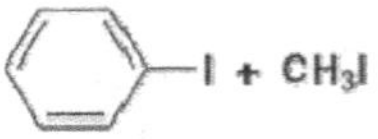 ⬡—I + CH$_3$I (d) 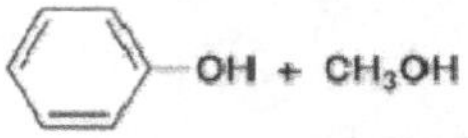 ⬡—OH + CH$_3$OH

5. Which cyclic alkene is the main product of acid-catalysed dehydration of *cis*-1,2-dimethyl cyclo hexanol?

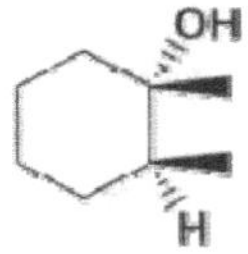

cis-1,2-dimethylcyclohexanol

(a) 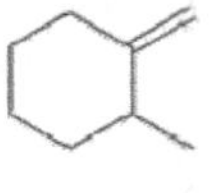 (b)

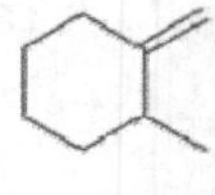

(c) 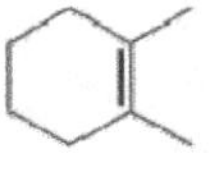 (d)

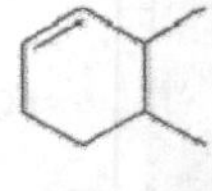

6. Which alkene is the main product of acid-catalysed dehydration of 3-methylbutan-1-ol? Note that the reaction may involve a rearrangement.

(a) 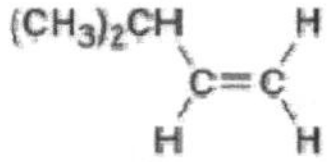 $(CH_3)_2CH$—CH=CH$_2$ (b) 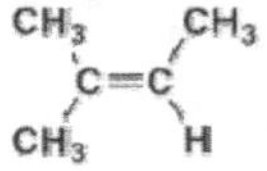 $(CH_3)_2C$=CH—CH$_3$

(c) 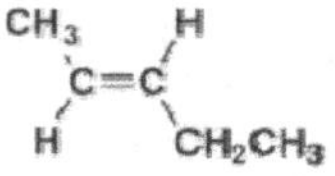 CH$_3$—CH=CH—CH$_2$CH$_3$ (d) 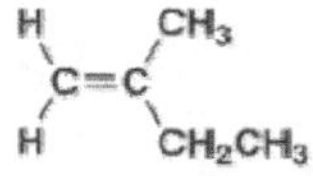 CH$_3$—CH=CH—CH$_2$CH$_3$

7. Which alcohol is the main product of base-catalysed reaction of epoxide 1 in methanol?

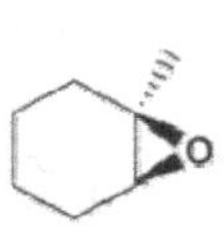

1

(a) 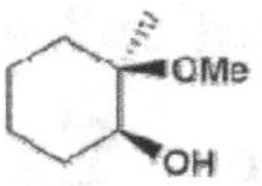 (b)

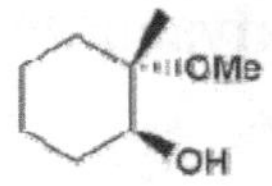

(c)   (d) — cyclohexane structures with OH and OMe substituents

8. Which of the following statements regarding sulfur compounds is false?

(a) A thiol is generally more acidic than an alcohol.

(b) A thiol is generally more nucleophilic than an alcohol.

(c) A thiolate can act as an oxidizing agent.

(d) A dialkyl sulfide can act as a nucleophile.

9. Which of the following statements regarding amines and ammonium compounds is false?

(a) Treatment of aliphatic primary amines with sodium nitrite in aqueous acid generates nitrogen.

(b) Treatment of aliphatic secondary amines with sodium nitrite in aqueous acid usually gives an oily product.

(c) Treatment of aliphatic tertiary amines with sodium nitrite in aqueous acid usually results in dissolution of the amine.

(d) Treatment of aliphatic quaternary ammonium compounds with sodium nitrite in aqueous acid usually gives an oily product.

10. Which of the following equations shows the correct final main products?

(a) $(CH_3)_2CHOCH_3 + HI \xrightarrow{H_2O} (CH_3)_2CHI + CH_3OH$

(b) $(CH_3)_3COCH_2CH_3 + HCl \xrightarrow{H_2O} (CH_3)_3COH + CH_3CH_2Cl$

(c) $CH_3CH_2\overset{\overset{+}{N}(CH_3)_3}{\underset{|}{CH}}CH_3$ + EtONa $\longrightarrow$ $CH_3CH_2CH{=}CH_2$ + $(CH_3)_3N$ + EtOH + NaI

(d) $CH_3CH_2NH_2$ + $NaNO_2$ + HCl $\xrightarrow{H_2O}$ $CH_3CH_2\underset{H}{N}{-}NO$ + H_2O + NaCl

11. Which compound is the most reactive in hydrochloric acid?

(a) 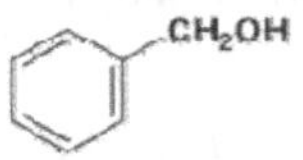   [benzyl alcohol, $C_6H_5{-}CH_2OH$]

(b)   [cyclohexylmethanol, $C_6H_{11}{-}CH_2OH$]

(c)   [1-phenylethanol, $C_6H_5{-}CH(CH_3){-}OH$]

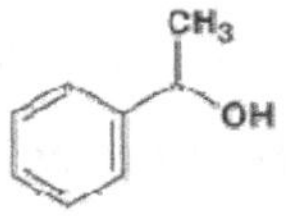

(d) 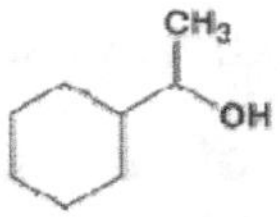   [1-cyclohexylethanol, $C_6H_{11}{-}CH(CH_3){-}OH$]

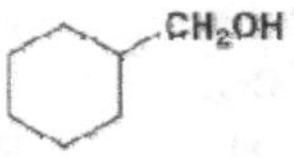

12. Which compound is the least reactive in hydrochloric acid?

(a) 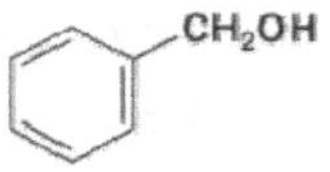   [benzyl alcohol, $C_6H_5{-}CH_2OH$]

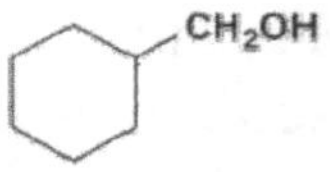

(b)   [cyclohexylmethanol, $C_6H_{11}{-}CH_2OH$]

(c)   [1-phenylethanol, $C_6H_5{-}CH(CH_3){-}OH$]

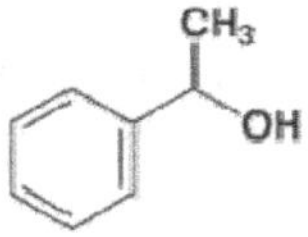

(d) 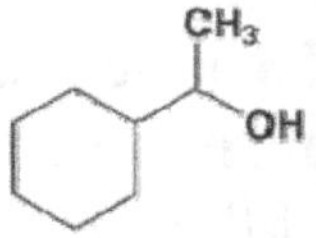   [1-cyclohexylethanol, $C_6H_{11}{-}CH(CH_3){-}OH$]

13. Which are the initial main products of the reaction of *t*-butyl ethyl ether in hydrochloric acid?

(a) $(CH_3)_3COH + CH_3CH_2Cl$

(b) $(CH_3)_3CCl + CH_3CH_2OH$

(c) $(CH_3)_3COH + CH_3CH_2OH$

(d) $(CH_3)_3CCl + CH_3CH_2Cl$

14. **Which are the initial main products of reaction of benzyl isopropyl ether in concentrated hydriodic acid?**

    (a) $PhCH_2OH + (CH_3)_2CHI$

    (b) $PhCH_2I + (CH_3)_2CHI$

    (c) $PhCH_2OH + (CH_3)_2CHOH$

    (d) $PhCH_2I + (CH_3)_2CHOH$

15. **Which alkene is the main product of the following reaction?**

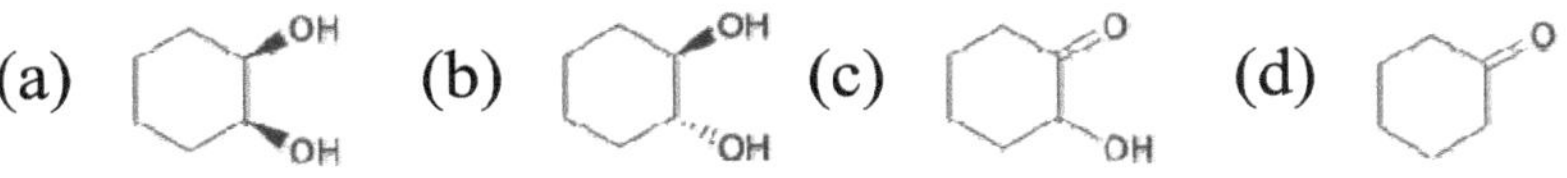

16. **Which carbenium ion is most stable?**

    (a) $(CH_3)_3C\overset{+}{C}H_2$

    (b) $(CH_3)_2\overset{+}{C}CH_2CH_3$

    (c) $(CH_3)_2CH\overset{+}{C}HCH_3$

    (d) $CH_3CH_2\overset{+}{C}HCH_2CH_3$

17. **Which carbenium ion is least stable?**

    (a) $(CH_3)_3C\overset{+}{C}H_2$

    (b) $(CH_3)_2\overset{+}{C}CH_2CH_3$

    (c) $(CH_3)_2CH\overset{+}{C}HCH_3$

    (d) $CH_3CH_2\overset{+}{C}HCH_2CH_3$

18. **Which of the following will not be formed in an aqueous acid-catalysed reaction of cyclohexene oxide?**

19. **Which of the following equations shows the correct main products?**

(a) C_6H_5—O—CH_2CH_3 + H_3O^+ I^- $\xrightarrow[-H_2O]{}$ C_6H_5—I + CH_3CH_2OH

(b) C_6H_5—CH_2—O—CH_2CH_3 + H_3O^+ Cl^- $\xrightarrow[-H_2O]{}$ C_6H_5—CH_2OH + CH_3CH_2Cl

(c) $\underset{\underset{CH_3CHCH_2CH_3}{|}}{OH}$ $\xrightarrow[\Delta]{cat.\ H_2SO_4}$ $CH_3CH{=}CHCH_3$ + H_2O

(d) $\underset{\underset{CH_3CHCH_2CH_3}{|}}{HO^-\ \overset{+}{N}(CH_3)_3}$ $\xrightarrow[\Delta]{}$ $CH_3CH{=}CHCH_3$ + H_2O + $N(CH_3)_3$

20. **Which of the following equations shows the wrong main product(s)?**

(a) $(CH_3CH_2)_2NH$ + CH_3CH_2I $\longrightarrow$ $(CH_3CH_2)_3\overset{+}{N}H\ I^-$

(b) $CH_3\overset{\overset{O}{\|}}{C}OEt$ + $HN{\bigcirc}$ $\longrightarrow$ $CH_3\overset{\overset{O}{\|}}{C}{-}N{\bigcirc}$ + $EtOH$

(c) $(CH_3)_2S$ + CH_3I $\longrightarrow$ $(CH_3)_3S^+\ I^-$

(d) $\underset{CH_3}{\triangle}\overset{O}{}$ + CH_3CH_2SNa $\xrightarrow[]{EtOH}$ $\underset{\underset{CH_3}{|}}{\overset{CH_3CH_2S}{\underset{}{}}}H{-}{\bigcirc}ONa$

<u>**KEYS**</u>

1. (c)	**2.** (d)	**3.** (b)	**4.** (a)	**5.** (a)
6. (a)	**7.** (a)	**8.** (a)	**9.** (a)	**10.** (a)
11. (a)	**12.** (a)	**13.** (b)	**14.** (c)	**15.** (a)
16. (b)	**17.** (b)	**18.** (b)	**19.** (c)	**20.** (d)

1. Which alkene is the most reactive in acid-catalysed hydration?

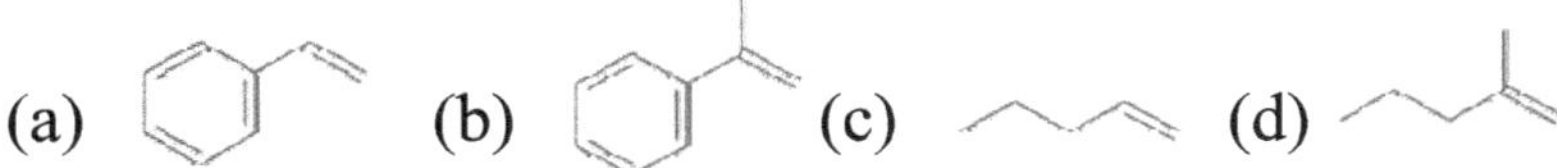

(a)　　(b)　　(c)　　(d)

2. Which alkene is the least reactive in acid-catalysed hydration?

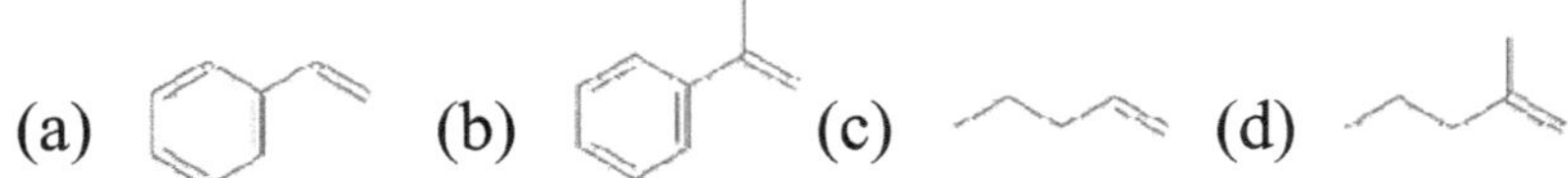

(a)　　(b)　　(c)　　(d)

3 Which of the following alkenes gives a main product different from the products from the others in reactions with hydrogen chloride?

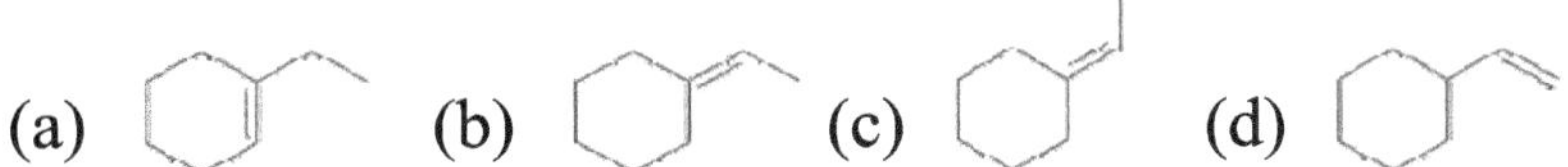

(a)　　(b)　　(c)　　(d)

4. Which is the main product when 2-methylbuta-1,3-diene reacts with one equivalent of HBr for a limited time below 0 °C?

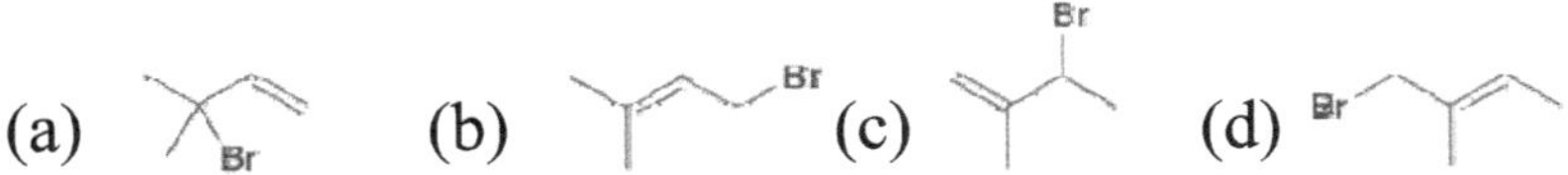

(a)　　(b)　　(c)　　(d)

5. **Which is the main product when 2-methylbuta-1,3-diene reacts with one equivalent of HBr at room temperature for a prolonged time?**

(a)   (b)   (c)   (d)

6. **Which alkene gives a *meso* compound upon reaction with Br_2 in CH_2Cl_2?**

(a)   (b)   (c)   (d)

7. **Which alkene gives a racemate upon reaction with Br_2 in CH_2Cl_2?**

(a)   (b)   (c)   (d)

8. **Which of the following reactions does not give 3-methylbutan-2-ol as the main product?**

(a) H_2SO_4/H_2O

(b) 1) $Hg(OAc)_2/H_2O$ 2) $NaBH_4/OH^-$

(c) 1) BH_3/THF 2) H_2O_2/OH^-

(d) 1) $Hg(OAc)_2/H_2O$ 2) $NaBH_4/OH^-$

9. **Which of the following equations does not show the correct main products?**

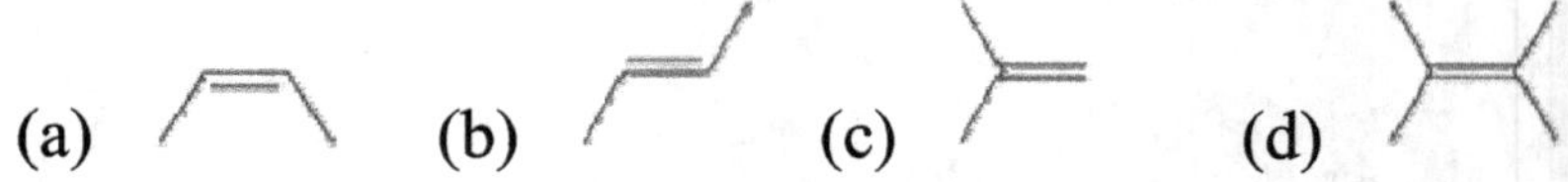

(a) $+ H_2O$ — H_2SO_4 →

(b) $+ HCl$ — $AcOH$ →

(c) 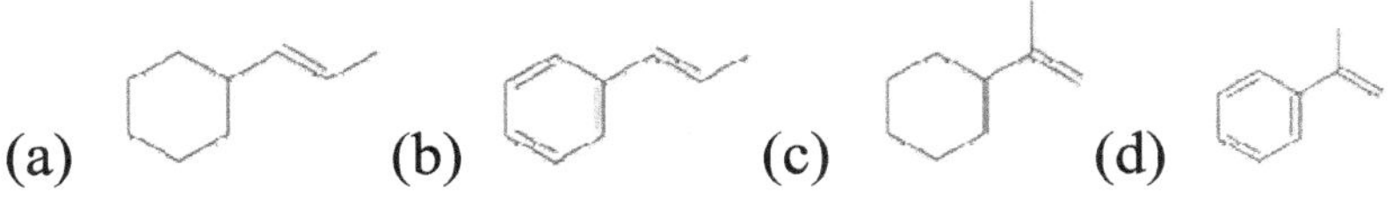

(d) 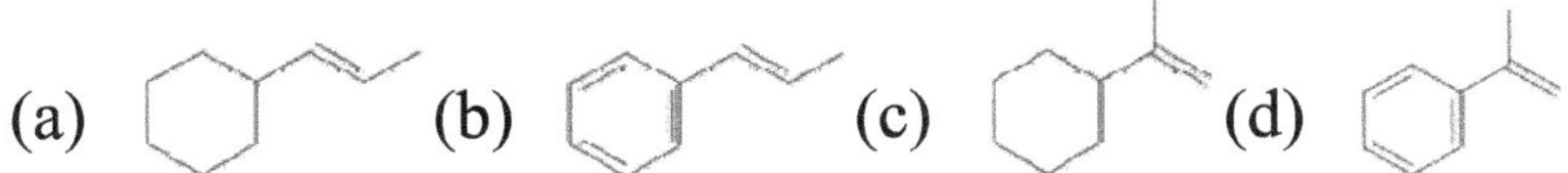

10. Which of the following statements is wrong?

(a) In the epoxidation of an alkene with a peroxy acid, the peroxy acid is electrophilic.

(b) The addition of bromine to cyclohexene is stereospecific but the product is a racemate.

(c) Hydroboration-oxidation of a terminal alkyne gives a ketone as the main product.

(d) Hydrogenation of an internal alkyne over the Lindlar catalyst gives a cis alkene.

11. Which alkene is most reactive in acid-catalysed hydration?

(a)   (b)   (c)   (d)

12. Which alkene is least reactive in acid-catalysed hydration?

(a)   (b)   (c)   (d)

13. Which alkene is most reactive in its reaction with hydrogen chloride?

(a)   (b)   (c)   (d)

14. Which alkene is most reactive in reaction with hydrogen chloride?

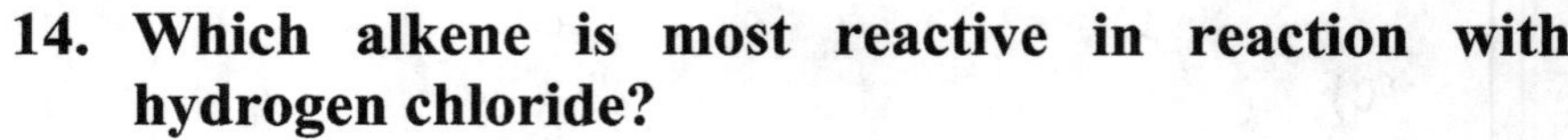

(a)   (b)   (c)   (d)

15. Which of the following alkenes gives a main product different from the products of the other three in reactions with hydrogen chloride?

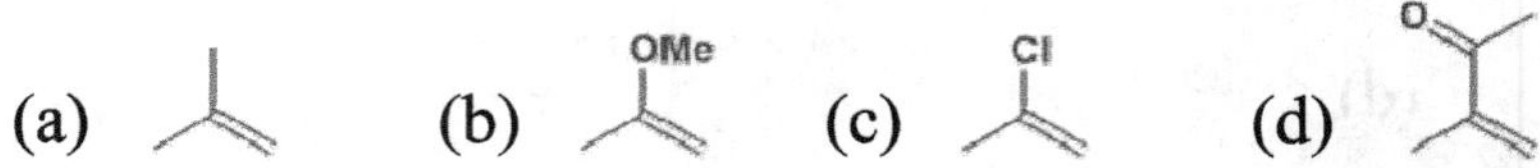

(a)   (b)   (c)   (d)

16. Which of the following reactions gives a *meso* compound as the main product?

(a) $\xrightarrow{\quad}$ + Br$_2$ $\xrightarrow{CH_2Cl_2}$

(b) + H$_2$ $\xrightarrow{Pt}$

(c) + Br$_2$ $\xrightarrow{CH_2Cl_2}$

(d) + H$_2$ $\xrightarrow{Pt}$

17. Which is the main product in the reaction of but-1-yne with an excess of hydrogen chloride?

(a) $CH_3CH_2\overset{Cl}{\underset{|}{C}}=CH_2$ (b) $CH_3CH_2CHCH_2Cl$ with Cl

(c) $CH_3CH_2\overset{Cl}{\underset{\underset{Cl}{|}}{\overset{|}{C}}}CH_3$ (d) $CH_3CH_2CH_2CHCl_2$

18. **Which is the main final product when the initial product (4,5-dibromopent-2-ene) from the 1,2-addition of bromine to penta-1,3-diene is heated in solution?**

(a) [structure: Br—CH2—CH=CH—CH(Br)—CH3]

(b) [structure: Br—CH2—CH2—CH=C(Br)—CH3]

(c) [structure: CH3—CH(Br)—C(Br)=CH—CH3]

(d) [structure: CH3—C(Br)=CH—CH(Br)—CH3]

19. **Which of the following equations does not show the correct main products?**

(a) HO—CH2—CH=CH2 $\xrightarrow[\text{2) H}_2\text{O}_2/\text{NaOH}]{\text{1) BH}_3/\text{THF}}$ HO—CH2—CH2—CH2—OH

(b) HO—CH2—CH=CH2 $\xrightarrow[\text{2) NaBH}_4/\text{NaOH}]{\text{1) Hg(OAc)}_2/\text{H}_2\text{O}}$ HO—CH2—CH2—CH2—OH

(c) HO—CH2—C≡CH $\xrightarrow{\text{Hg(OAc)}_2/\text{H}_2\text{O}}$ HO—CH2—C(=O)—CH3

(d) HO—CH2—CH2—C≡CH $\xrightarrow[\text{2) H}_2\text{O}_2/\text{NaOH}]{\text{1) BH}_3/\text{THF}}$ HO—CH2—CH2—CH2—CHO

20. **Which of the following statements is wrong?**

(a) The main orbital interactions in the transition state of an electrophilic addition to an alkene are between the LUMO of the electrophile and the HOMO of the alkene.

(b) The lower the LUMO of the electrophile, the more reactive it is.

(c) The lower the HOMO of the alkene, the more reactive it is.

(d) The electrophile approaches the alkene from above its molecular plane in addition reactions.

<u>KEYS</u>

1. (b)	**2.** (c)	**3.** (d)	**4.** (a)	**5.** (b)
6. (b)	**7.** (a)	**8.** (d)	**9.** (b)	**10.** (c)
11. (d)	**12.** (a)	**13.** (d)	**14.** (b)	**15.** (c)
16. (a)	**17.** (c)	**18.** (a)	**19.** (b)	**20.** (c)

ELECTROPHILLIC AROMATIC SUBSTITUTION

1. **Which is most reactive in electrophilic substitution?**

(a) 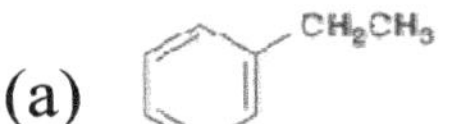 CH_2CH_3

(b) OCH_3

(c) $\overset{O}{\overset{\|}{C}}CH_3$

(d) Cl

2. **Which is least reactive in electrophilic substitution?**

(a) CH_2CH_3

(b) OCH_3

(c) $\overset{O}{\overset{\|}{C}}CH_3$

(d) Cl

3. **Which gives a *meta* nitro compound as the main product upon nitration with a nitric acid-sulfuric acid mixture?**

(a) CH_2CH_3

(b) OCH_3

(c) $\overset{O}{\overset{\|}{C}}CH_3$

(d) Cl

4. Which is most reactive in electrophilic substitution?

(a) 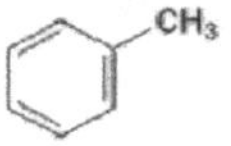

(b) 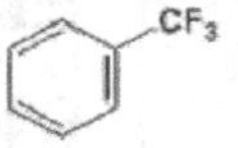

(c) 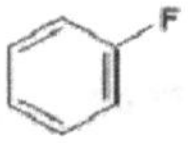

(d) 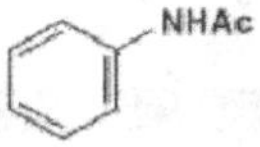

5. Which is least reactive in electrophilic substitution?

(a) 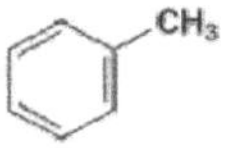

(b) 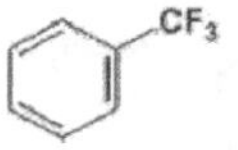

(c) 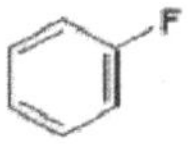

(d) 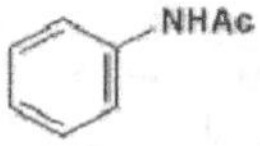

6. Which gives a *meta* nitro compound as the main product upon nitration with a nitric acid-sulfuric acid mixture?

(a) 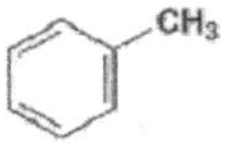

(b) 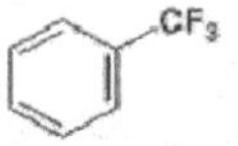

(c) 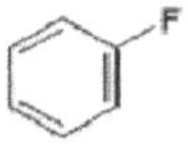

(d) 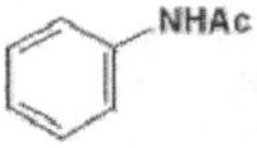

7. Which will be the main product upon chlorination of *m*-nitrotoluene with $Cl_2/AlCl_3$?

(a) 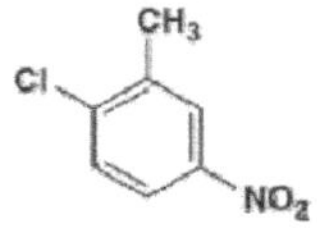

(b) 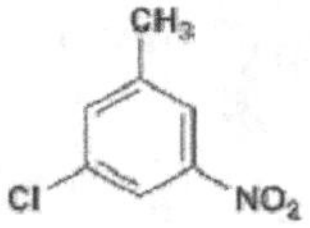

(c) 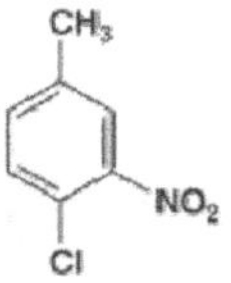

(d) 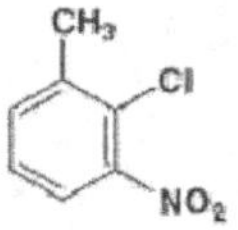

8. **Which is obtained as the main mononitration product upon reaction of *m-t*-butylanisole (1-*t*-butyl-3-methoxybenzene) with HNO$_3$-H$_2$SO$_4$?**

(a) 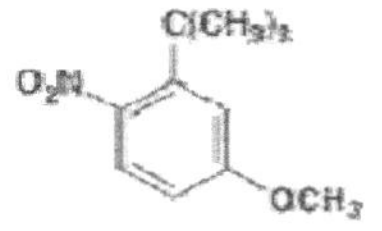 [structure: benzene ring with C(CH$_3$)$_3$, O$_2$N, OCH$_3$]

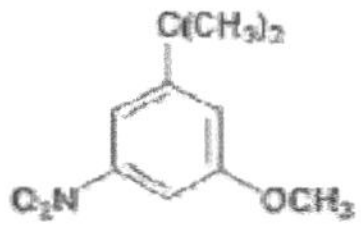

(b) [structure: benzene ring with C(CH$_3$)$_3$, O$_2$N, OCH$_3$]

(c) [structure: benzene ring with C(CH$_3$)$_3$, OCH$_3$, NO$_2$]

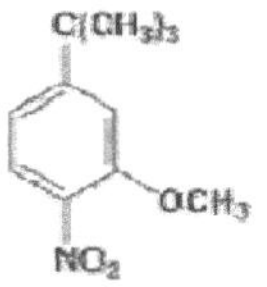

(d) 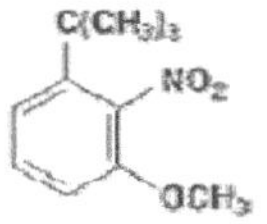 [structure: benzene ring with C(CH$_3$)$_3$, NO$_2$, OCH$_3$]

9. **Which of the following statements regarding Friedel-Crafts reactions is wrong?**

(a) Alkylation of benz ene w ith an a lkyl chlori de requires only a cata lytic amount of a L ewis aci d such as aluminum chloride.

(b) b)Alkylation of benz ene w ith an a lcohol requires only a catalytic am ount of a Brønsted acid such as phosphoric acid.

(c) Acetylation of be nzene with a cetyl chlori de requires only a catalytic amount of a Lewis acid.

(d) Acetylation of benzene w ith ace tic anhy dride requires m ore tha n one equivalent of a Lew is acid.

10. **Which of the following statements regarding electrophilic aromatic substitution is wrong?**

(a) Acetyl and cyano substituents are both deactivating and *m*-directing.

(b) Alkyl groups are activating and *o,p*-directing.

(c) Ammonio groups are *m*-directing but amino groups are and *o,p*-directing.

(d) Chloro and methoxy substituents are both deactivating and *o,p*-directing.

11. Which is most reactive in electrophilic substitution?

(a) 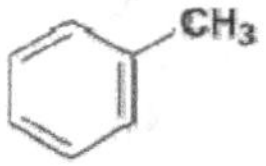 [benzene ring with CH₃]

(b) 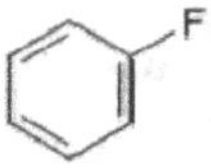 [benzene ring with F]

(c) 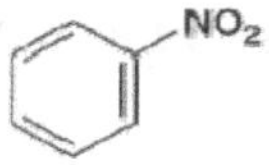 [benzene ring with NO₂]

(d) 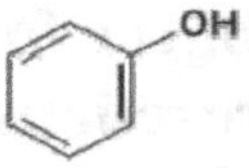 [benzene ring with OH]

12. Which is least reactive in electrophilic substitution?

(a) 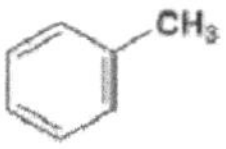 [benzene ring with CH₃]

(b) 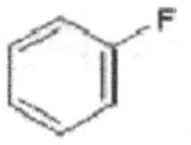 [benzene ring with F]

(c) 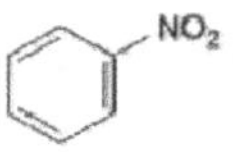 [benzene ring with NO₂]

(d) 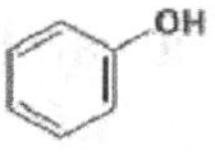 [benzene ring with OH]

13. Which gives a *para* nitro compound as the main product upon nitration with a nitric acid-sulfuric acid mixture?

(a) 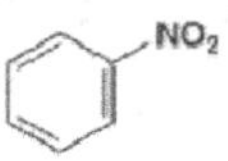 [benzene ring with NO₂]

(b) 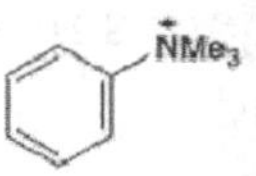 [benzene ring with $\overset{+}{N}Me_3$]

(c) 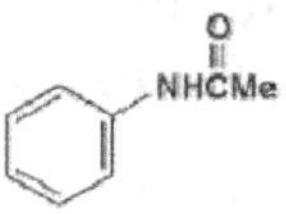 [benzene ring with NHCMe, C=O]

(d) 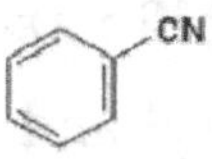 [benzene ring with CN]

14. Which gives a *meta* nitro compound as the main product upon nitration with a nitric acid-sulfuric acid mixture?

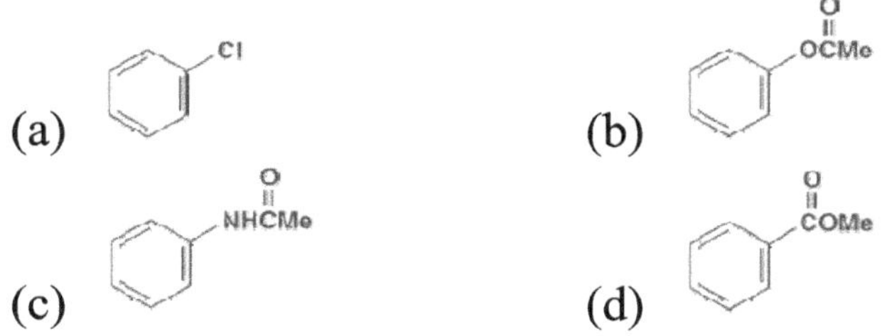

(a)   (b)   (c)   (d)

15. Which of (a)-(d) does not give isopropylbenzene as a product upon reaction with benzene?

(a) $(CH_3)_2CHCl/AlCl_3$ (b) $CH_3CH_2CH_2Cl/AlCl_3$

(c) $CH_3CH=CH_2/H_3PO_4$ (d) $(CH_3)_2C=CH_2/H_3PO_4$

16. Which combination of reagents used in the indicated order with benzene will give *m*-nitropropylbenzene?

(a) 1. HNO_3/H_2SO_4,

     2. $H_3CH_2CH_2Cl/AlCl_3$

(b) 1. $CH_3CH_2CH_2Cl/AlCl_3$,

     2. HNO_3/H_2SO_4

(c) 1. $CH_3CH_2COCl/AlCl_3$,

     2. HNO_3/H_2SO_4,

     3. $H_2NNH_2/NaOH$

(d) 1. HNO_3/H_2SO_4,

     2. $CH_3CH_2COCl/AlCl_3$,

     3. $H_2NNH_2/NaOH$

17. Which pair of compounds are the most probable main products of the following reaction?

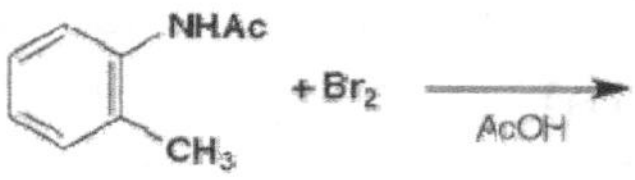

(a) 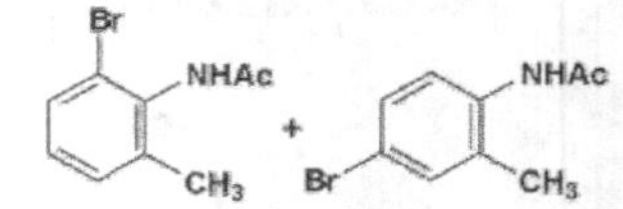

(b) 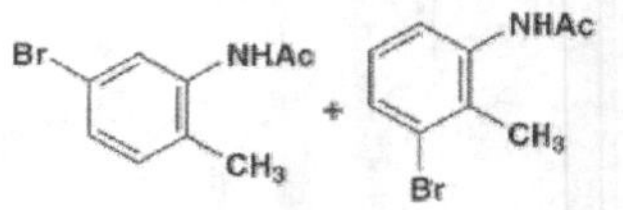

(c) 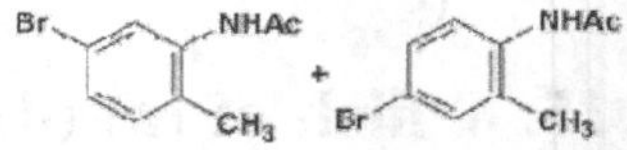

(d) 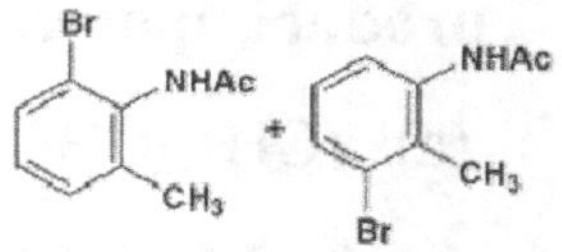

18. Which pair of compounds are the most probable main products of the following reaction?

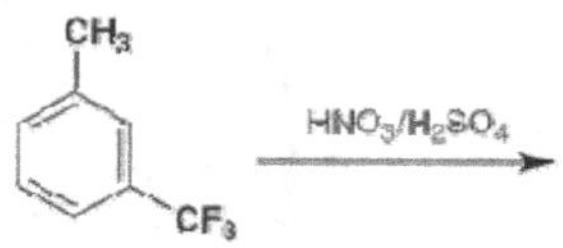

(a) 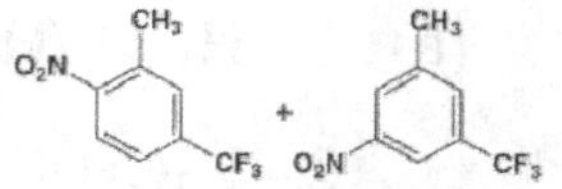

(b) 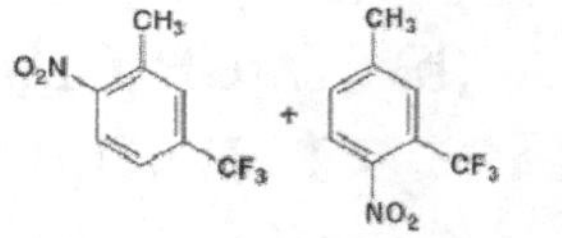

(c) 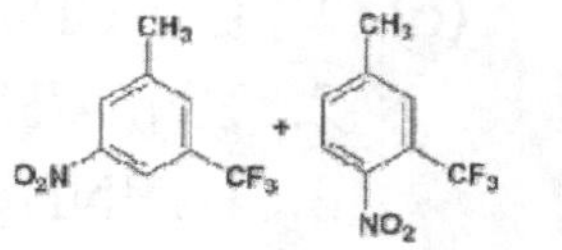

(d) 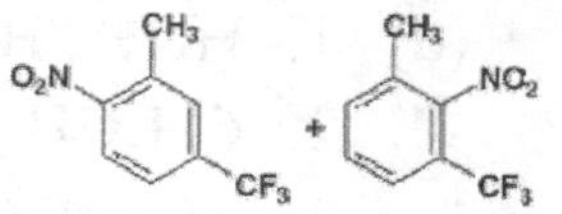

19. **Which will be the main product upon bromination of phenyl benzoate with Br$_2$/AlBr$_3$?**

(a)

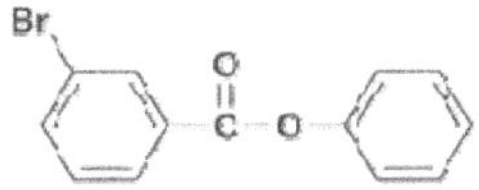

(b)

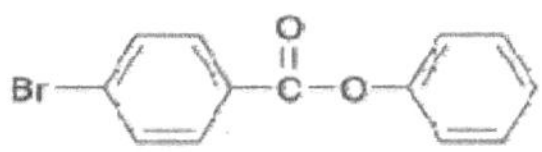

(c)

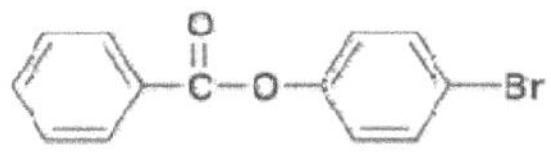

(d)

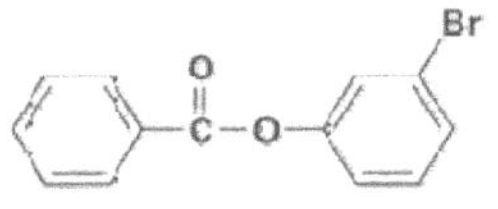

20. **Which of the following statements regarding electrophilic aromatic substitution is wrong?**

(a) Sulfonation of toluene is reversible.

(b) Friedel-Crafts alkylation of benzene can be reversible.

(c) Friedel-Crafts alkylation with primary alkyl chloride may involve rearrangement.

(d) Friedel-Crafts acylation of nitrobenzene readily gives a meta substitution product.

KEYS

1. (b)	**2.** (c)	**3.** (c)	**4.** (d)	**5.** (d)
6. (b)	**7.** (a)	**8.** (c)	**9.** (c)	**10.** (d)
11. (d)	**12.** (c)	**13.** (b)	**14.** (c)	**15.** (a)
16. (b)	**17.** (b)	**18.** (b)	**19.** (c)	**20.** (d)

CHAPTER 15 POLYCYCLIC AND HETEROCYCLIC AROMATIC COMPOUNDS

1. **Which is the most probable main product of the following reaction**

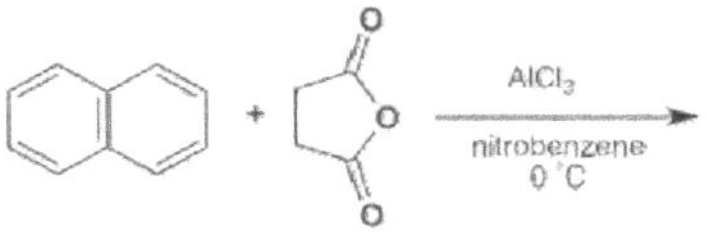

(a) 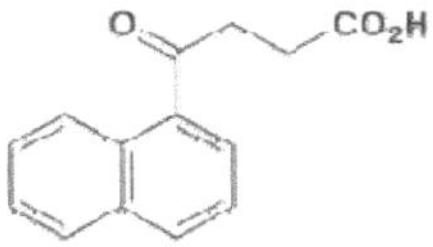

(b) 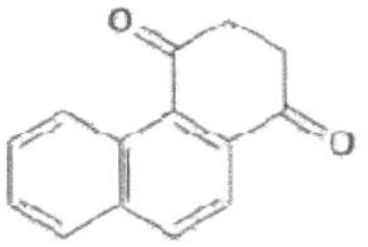

(c) 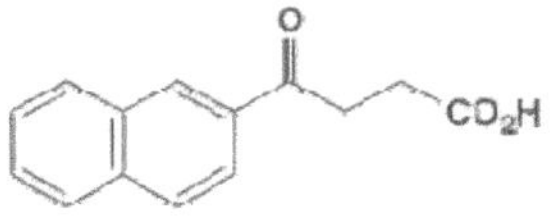

(d) 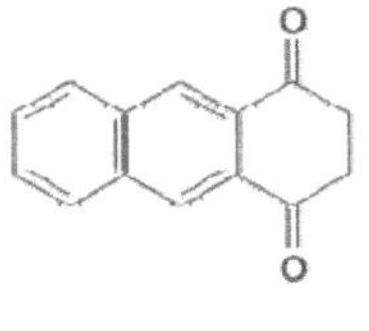

2. **Which is the most probable main product of the following reaction?**

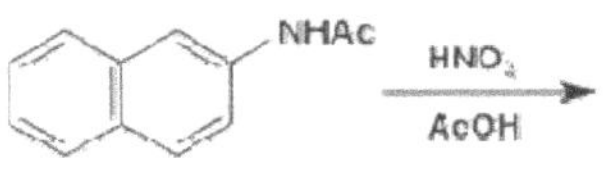

(a) 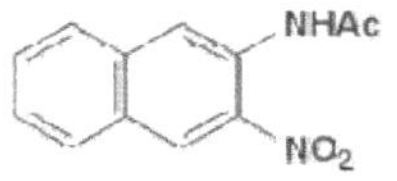

(b) 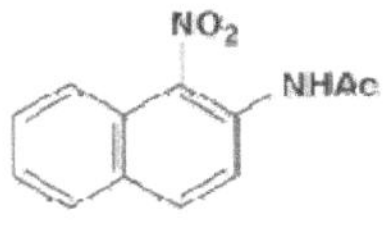

(c)

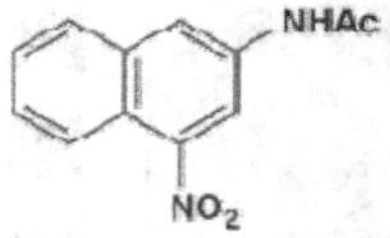

(d)

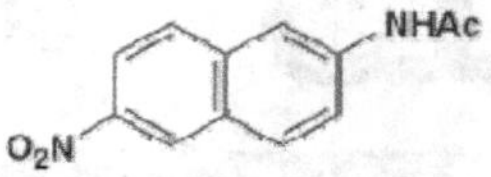

3. Which is the most probable main product of the following reaction?

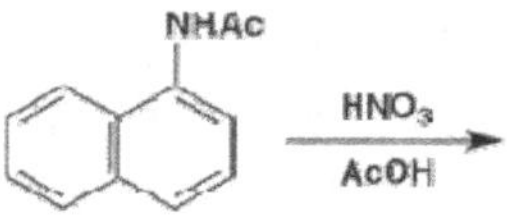

(a)

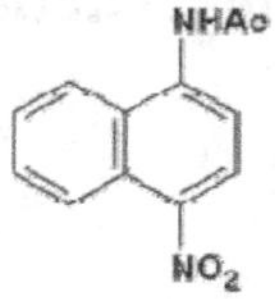

(b)

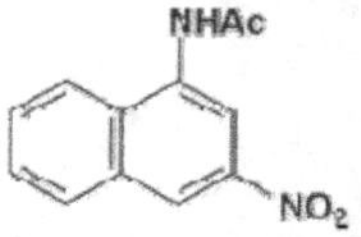

(c)

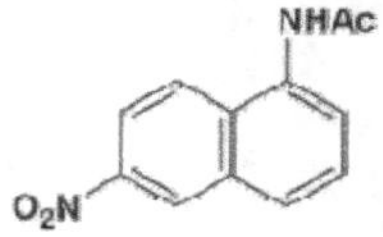

(d)

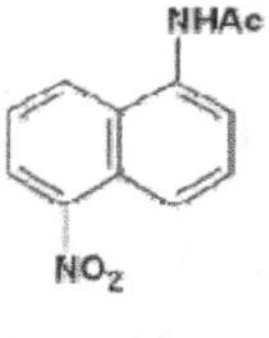

4. Which is the most probable main product of the following reaction?

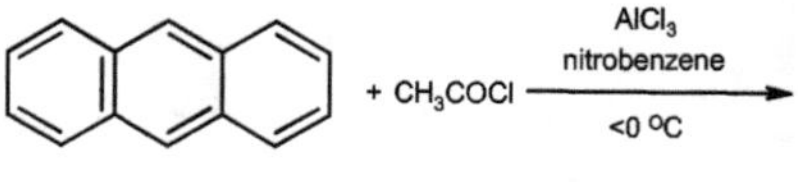

(a)

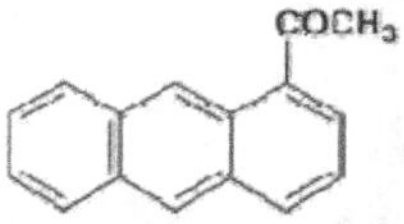

(b)

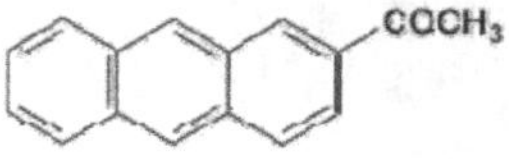

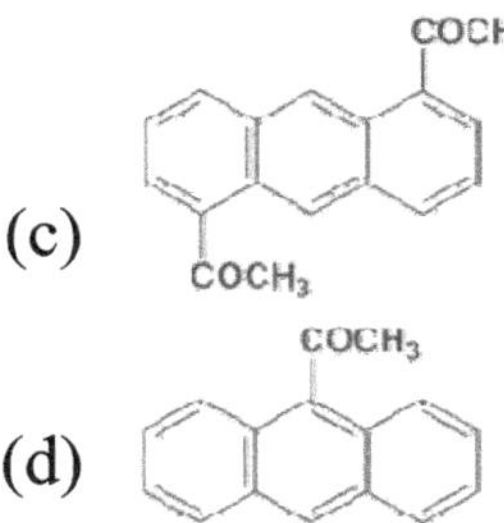

(c)

(d)

5. Which compound is most basic?

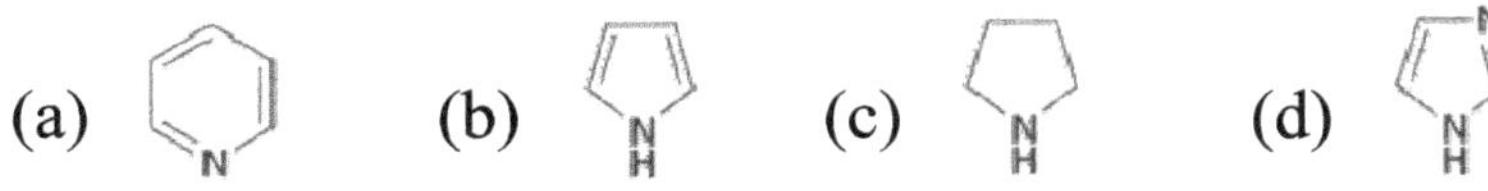

(a)　　　(b)　　　(c)　　　(d)

6. Which compound is least basic?

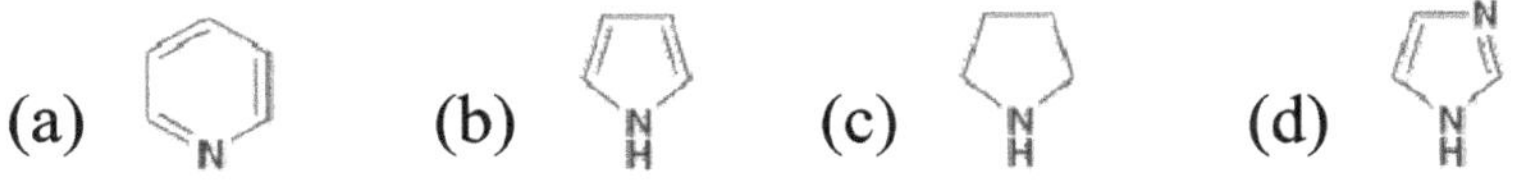

(a)　　　(b)　　　(c)　　　(d)

7. Which is most reactive in electrophilic aromatic substitution?

(a)　　　(b)　　　(c)　　　(d)

8. Which of (a)-(d) is not aromatic?

(a)　　　　　　(b)

(c)　　　　　　(d)

9. Which is the most probable main product of the following reaction?

(a)

(b)

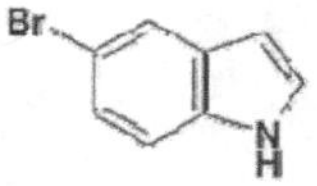

(c)

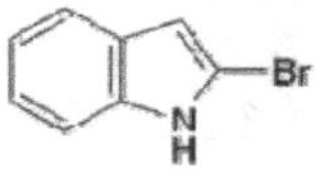

(d)

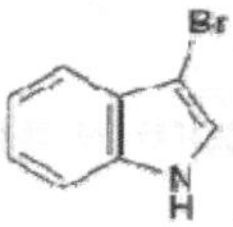

10. Which of the following equations shows an unlikely result?

(a)

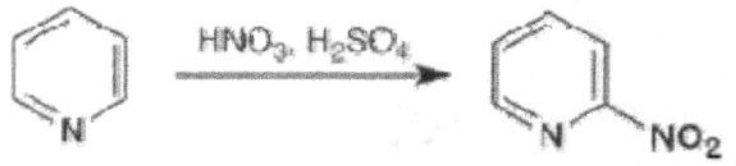

(b)

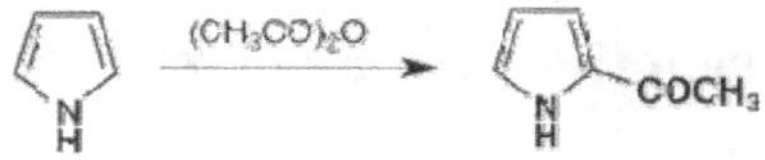

(c)

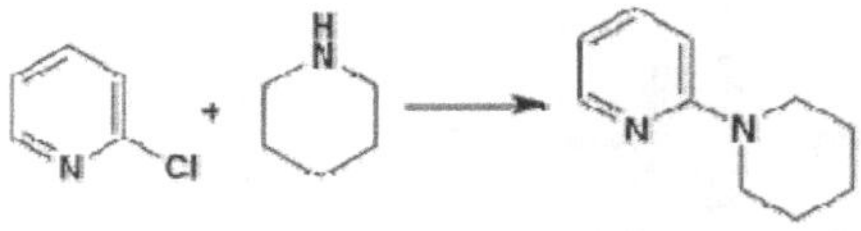

(d)

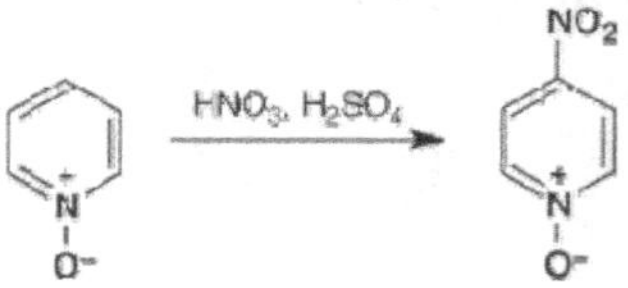

11. Which is impossible as a resonance contributor of anthracene.

(a) 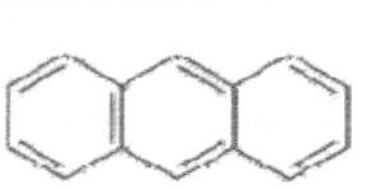   (b)

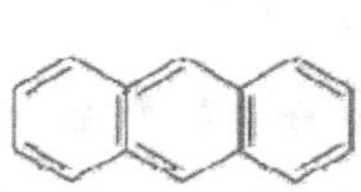

(c) 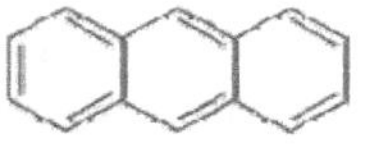   (d)

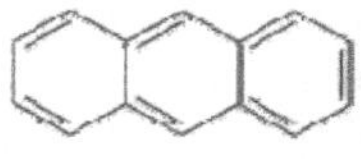

12. **Which of (a)-(d) is the most important resonance contributor of the intermediate naphthalenium ion in an electrophilic substitution reaction of naphthalene?**

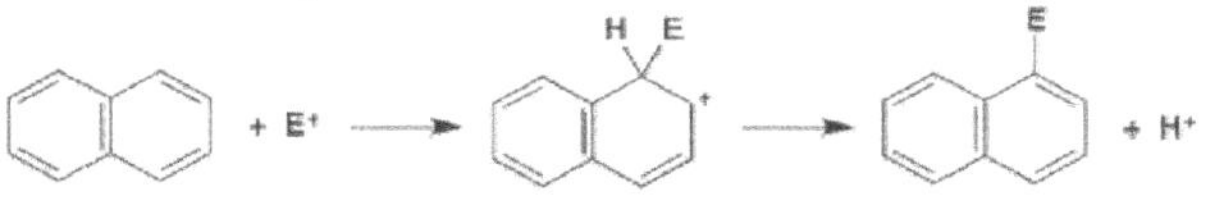

(a)

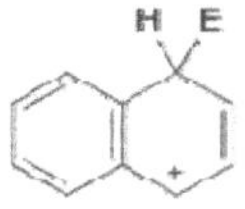

(b)

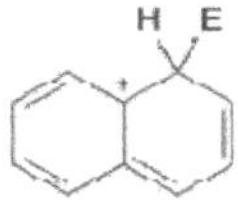

(c)

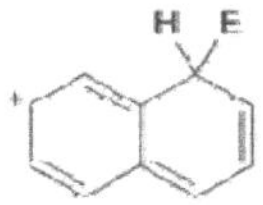

(d)

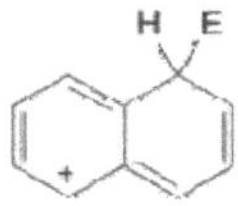

13. **Naphthalene-2-ol (2-naphthol) readily gives a dibromo substitution product with bromine in ethanoic acid. What is the most likely structure of this compound?**

a)

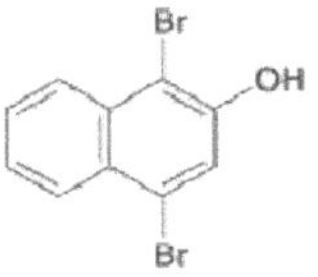

(b)

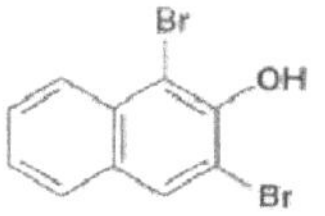

c)

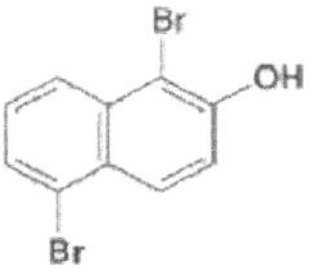

(d)

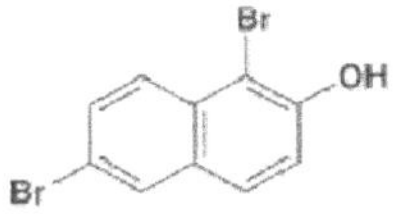

14. **Which is the most probable main product of the following reaction?**

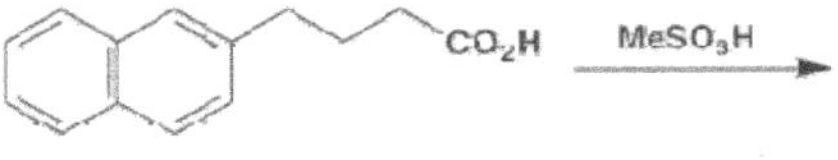

(a)

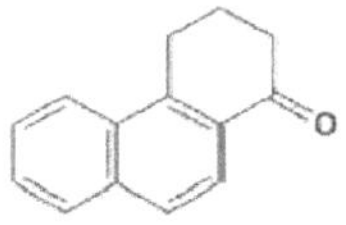

(b)

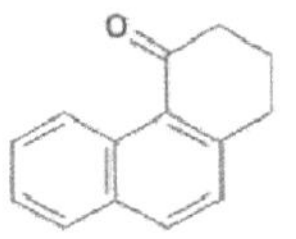

(c) 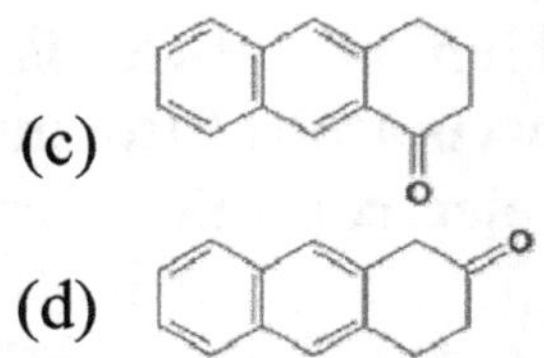

(d)

15. Which is most reactive towards an electrophile?

(a)   (b)   (c)   (d)

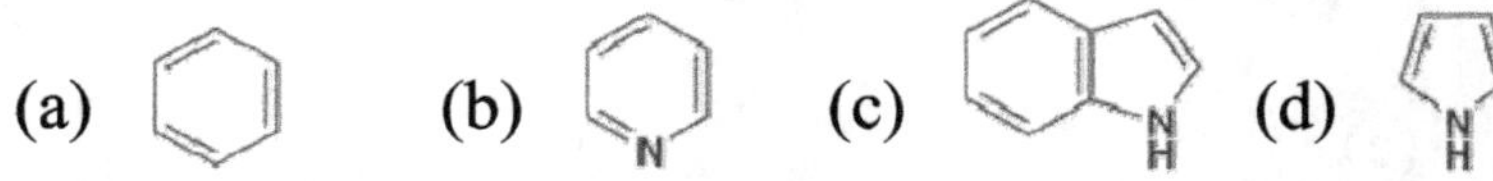

16. Which is least reactive towards an electrophile?

(a)   (b)   (c)   (d)

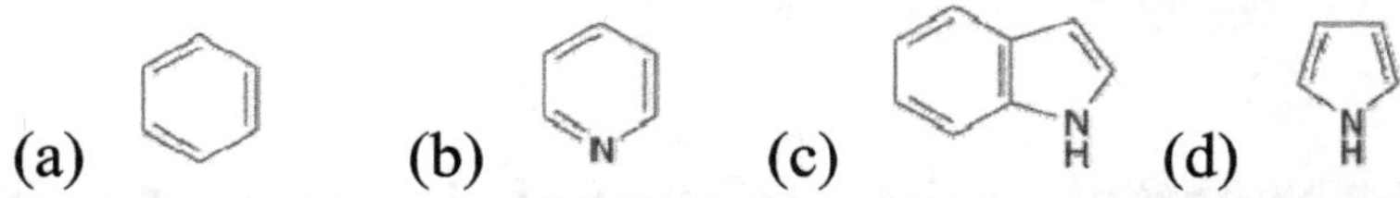

17. Which is most stabilized by electron delocalization (resonance)?

(a)   (b)   (c)   (d)

18. Which of (a)-(d) is not aromatic?

(a)   (b)   (c)   (d)

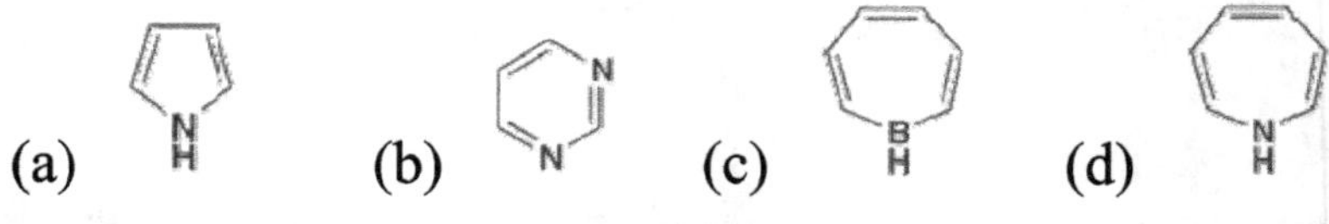

19. Which is the most probable main product of the following reaction?

(a)

(b)

(c)

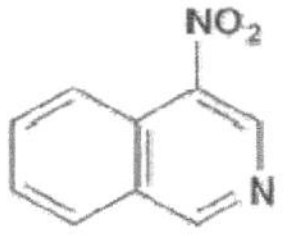

(d)

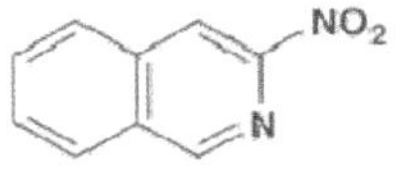

20. Which of the following reactions does not give a heterocyclic compound as indicated?

(a)

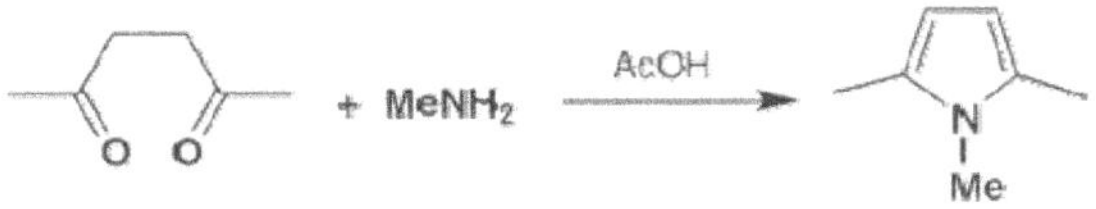

(b)

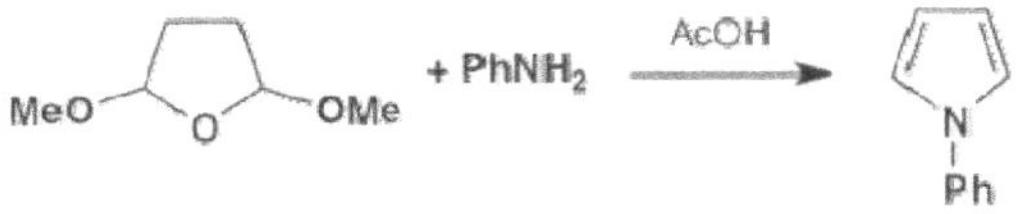

(c)

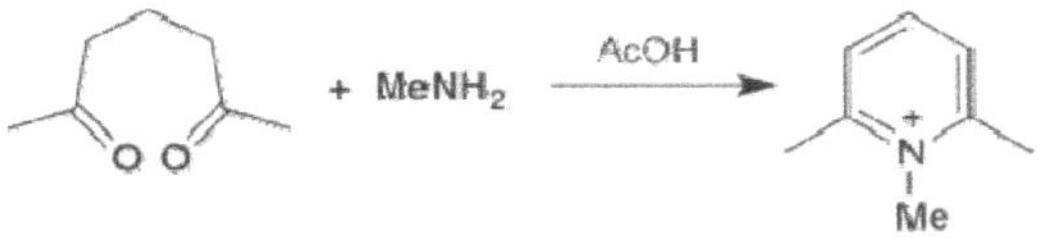

(d)

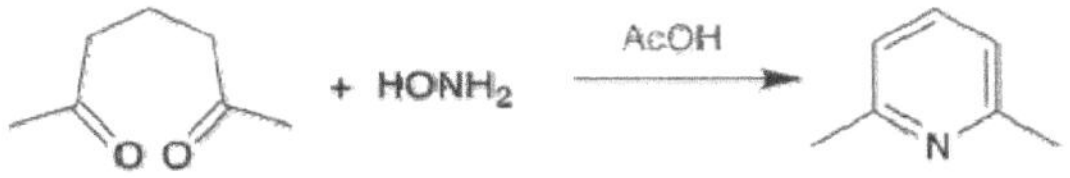

<u>**KEYS**</u>

1. (a)	**2.** (b)	**3.** (a)	**4.** (d)	**5.** (c)
6. (b)	**7.** (b)	**8.** (c)	**9.** (d)	**10.** (a)
11. (b)	**12.** (a)	**13.** (d)	**14.** (b)	**15.** (d)
16. (b)	**17.** (c)	**18.** (d)	**19.** (a)	**20.** (c)

1. **Which of the following dienophiles is the most reactive with buta-1,3-diene?**

   (a) $CH_2=CH-CO_2Me$

   (b) $CH_2=CH-OEt$

   (c) maleic anhydride

   (d) $MeO_2C-CH=CH-CO_2Me$

2. **Which of the following dienes cannot undergo Diels-Alder reactions?**

   (a) 2-methylbuta-1,3-diene (isoprene)

   (b) hexa-2,4-diene

   (c) benzene

   (d) methylenecyclohexene

3. **Which of adducts (a)-(d) is the main product of the following Diels-Alder reaction?**

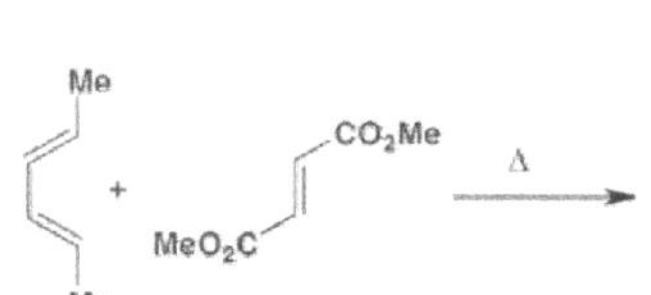

   (a)

   (b)

   (c)

   (d)

4. Which of the following is the main product in the ozonolysis of 1,4-dimethylcyclohexene followed by a reductive workup with Zn and ethanoic acid?

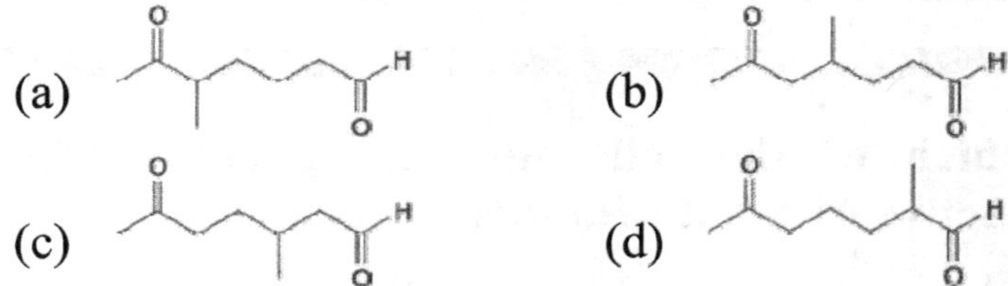

(a)　　　　　　　　　　　　(b)

(c)　　　　　　　　　　　　(d)

5. Which of the following is the main product in the ozonolysis of 1,5-dimethylcyclohexene followed by a workup with aqueous H_2O_2?

(a)　　　　　　　　　　　　(b)

(c)　　　　　　　　　　　　(d)

6. Which of phenols (a)-(d) is the main product of the following thermal rearrangement?

(a)

(b)

(c)

(d)

7. **Which of unsaturated aldehydes (a)-(d) is the sigmatropic rearrangement product obtained by heating the following ether?**

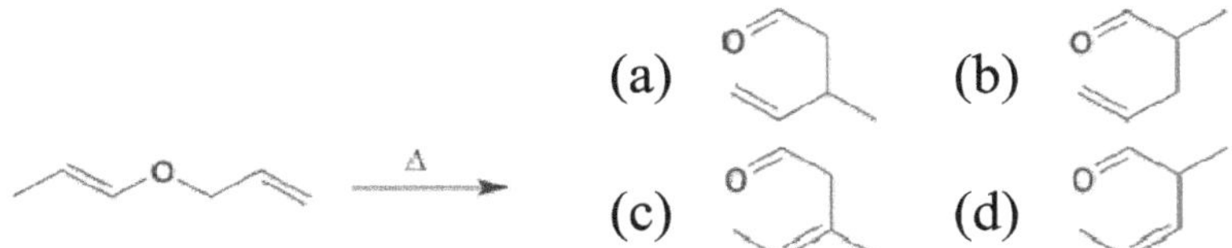

8. **Which of the following reactions is not thermally allowed?**

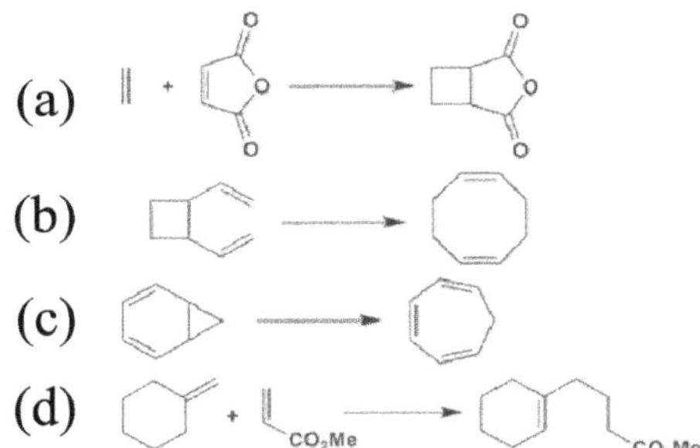

9. **Which of the following reactions is classified as an ene reaction?**

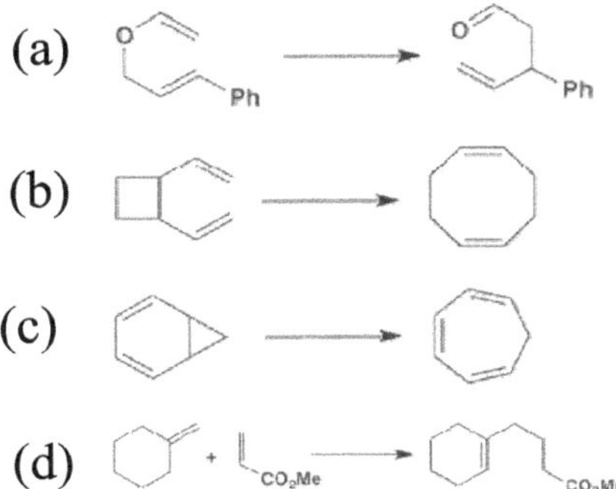

10. Which of the following reactions is classified as an electrocyclic reaction?

(a)

(b)

(c)

(d)

11. Which of the following dienophiles is the most reactive in normal Diels-Alder reactions?

(a) COMe

(b) OCOMe

(c) CO_2Me

(d) Me CO_2Me

12. Which of the following dienophiles is the least reactive in normal Diels-Alder reactions?

(a) COMe

(b) OCOMe

(c) CO_2Me

(d) Me CO_2Me

13. Which of adducts (a)-(d) is the main kinetic product of the following Diels-Alder reaction?

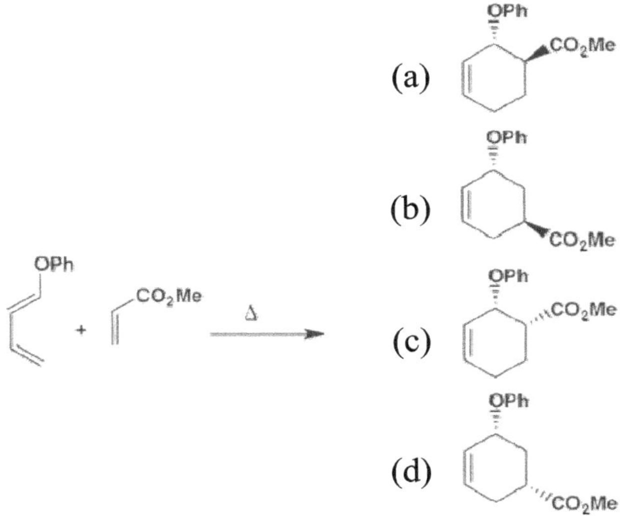

14. Which of the following is the main product in the ozonolysis of 1,3-dimethylcyclohexene followed by a workup with Zn and ethanoic acid?

(a)   (b)

c)   (d)

15. Which of the following is impossible as a resonance form of ozone?

(a)   (b)

(c)   (d)

16. **Which side-chain carbon makes a new bond to the benzene ring upon Claisen rearrangement of the following allylic phenyl ether?**

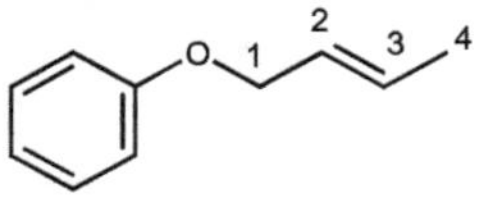

   (a) C1
   (b) C2
   (c) C3
   (d) C4

17. **The following involves two pericyclic reactions. Which combination indicates correctly the types of reaction involved?**

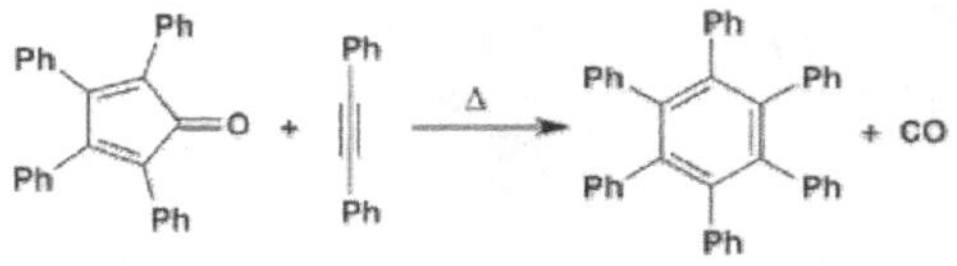

   (a) [4+2] cycloaddition + [2+2] cycloreversion
   (b) cheletropic reaction + [4+2] cycloaddition
   (c) [4+2] cycloaddition + [4+1] cycloreversion
   (d) [4+2] cycloaddition + cheletropic reaction

18. **Which of compounds (a)-(d) is the likely intermediate of the following transformation involving two electrocyclic reactions (ring opening and ring closure)?**

   (a)

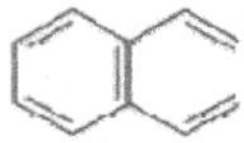

   (b)

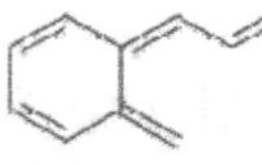

   (c)

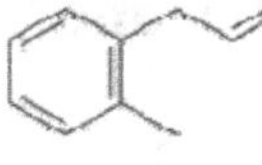

   (d)

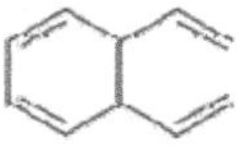

19. Which of the following reactions is classified as a sigmatropic rearrangement?

(a)

(b)

(c)

(d)

20. Which of the following is classified as an electrocyclic reaction?

(a)

(b)

(c)

(d)

<u>**KEYS**</u>

1. (c)	**2.** (d)	**3.** (b)	**4.** (c)	**5.** (c)
6. (d)	**7.** (b)	**8.** (a)	**9.** (d)	**10.** (c)
11. (a)	**12.** (b)	**13.** (c)	**14.** (d)	**15.** (c)
16. (c)	**17.** (d)	**18.** (b)	**19.** (a)	**20.** (b)

REARRANGEMENT REACTIONS

1. **Which is an unlikely product of the following hydrochlorination of an alkene?**

$$H-\underset{\underset{CH_3}{|}}{\overset{\overset{CH_3}{|}}{C}}-CH=CHCH_3 \;+\; HCl \;\longrightarrow$$

(a) $H-\underset{\underset{CH_3}{|}}{\overset{\overset{CH_3}{|}}{C}}-\underset{\underset{H}{|}}{\overset{\overset{Cl}{|}}{C}}-CH_2CH_3$

(b) $H-\underset{\underset{CH_3}{|}}{\overset{\overset{CH_3}{|}}{C}}-CH_2-\underset{\underset{H}{|}}{\overset{\overset{Cl}{|}}{C}}-CH_3$

(c) $CH_3-\underset{\underset{Cl}{|}}{\overset{\overset{CH_3}{|}}{C}}-\underset{\underset{H}{|}}{\overset{\overset{H}{|}}{C}}-CH_2CH_3$

(d) $H-\underset{\underset{CH_3}{|}}{\overset{\overset{CH_3}{|}}{C}}-CH_2-CH_2-CH_2Cl$

2. **Which of cations (a)-(d) is the least likely to be an intermediate in the following rearrangement reaction?**

$$CH_3-\underset{\underset{OH}{|}}{\overset{\overset{CH_3}{|}}{C}}-\underset{\underset{OH}{|}}{\overset{\overset{CH_3}{|}}{C}}-CH_3 \;\xrightarrow{H^+}\; CH_3-\underset{\overset{||}{O}}{\overset{}{C}}-\underset{\underset{CH_3}{|}}{\overset{\overset{CH_3}{|}}{C}}-CH_3$$

(a) $CH_3-\underset{\underset{OH}{|}}{\overset{\overset{CH_3}{|}}{C}}-\underset{\underset{^+OH_2}{|}}{\overset{\overset{CH_3}{|}}{C}}-CH_3$

(b) $CH_3-\underset{\underset{^+OH_2}{|}}{\overset{\overset{CH_3}{|}}{C}}-\underset{\underset{^+OH_2}{|}}{\overset{\overset{CH_3}{|}}{C}}-CH_3$

(c) 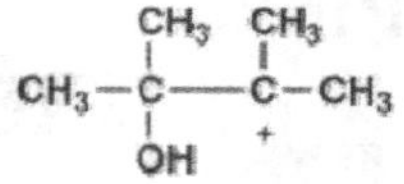 $CH_3-\overset{\overset{\displaystyle CH_3}{|}}{\underset{\underset{\displaystyle OH}{|}}{C}}-\overset{\overset{\displaystyle CH_3}{|}}{\underset{+}{C}}-CH_3$

(d) 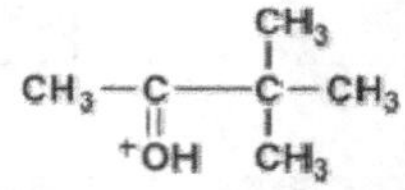 $CH_3-\overset{}{\underset{\underset{\displaystyle {}^+OH}{||}}{C}}-\overset{\overset{\displaystyle CH_3}{|}}{\underset{\underset{\displaystyle CH_3}{|}}{C}}-CH_3$

3. **Which cation is not involved in the following rearrangement of an epoxide?**

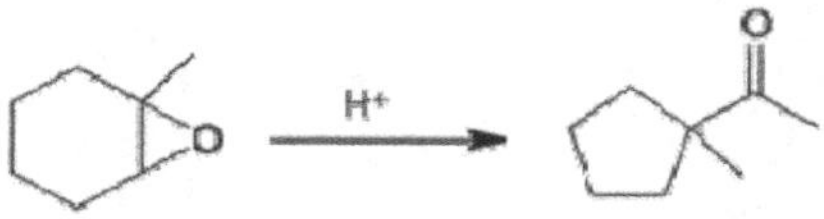

(a)

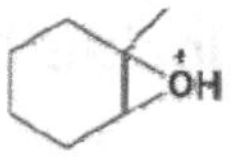

(b)

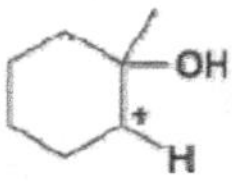

(c)

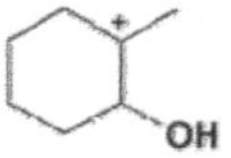

(d)

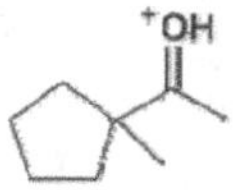

4. **Which carbonyl compound is the main product of the following reaction of a diol?**

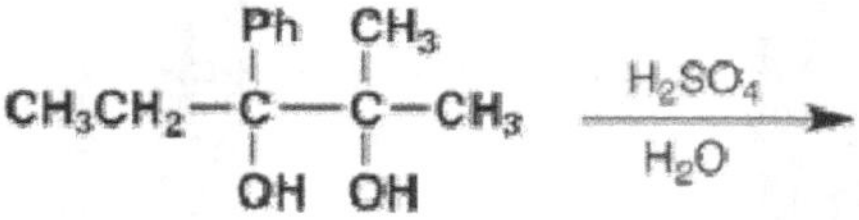

(a) 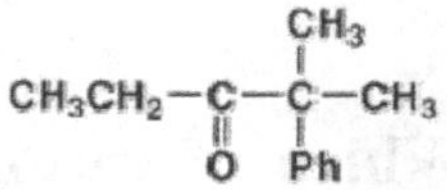 $CH_3CH_2-\overset{}{\underset{\underset{\displaystyle O}{||}}{C}}-\overset{\overset{\displaystyle CH_3}{|}}{\underset{\underset{\displaystyle Ph}{|}}{C}}-CH_3$

(b) $Ph-\overset{}{\underset{\underset{\displaystyle O}{||}}{C}}-\overset{\overset{\displaystyle CH_3}{|}}{\underset{\underset{\displaystyle CH_2CH_3}{|}}{C}}-CH_3$

(c) $CH_3-\overset{\overset{\displaystyle Ph}{|}}{\underset{\underset{\displaystyle CH_3}{|}}{C}}-\overset{}{\underset{\underset{\displaystyle O}{||}}{C}}-CH_2CH_3$

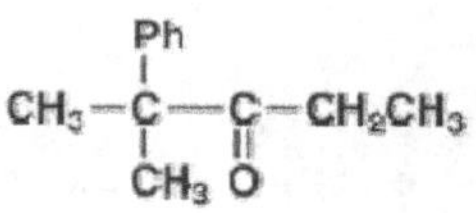

(d) 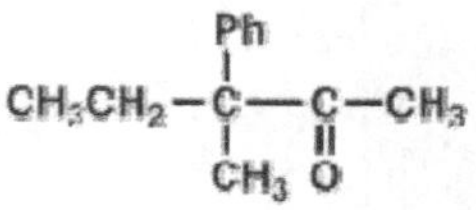 $CH_3CH_2-\overset{\overset{\displaystyle Ph}{|}}{\underset{\underset{\displaystyle CH_3}{|}}{C}}-\overset{}{\underset{\underset{\displaystyle O}{||}}{C}}-CH_3$

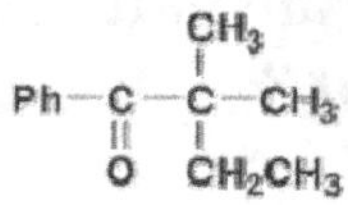

5. Which carbonyl compound is the main product of the following reaction of a tosylate?

$$CH_3CH_2-\underset{OTs}{C}(H)-\underset{OH}{C}(CH_3)-CH_3 \xrightarrow{H_2O}$$

(a) $CH_3CH_2-\overset{O}{C}-\overset{CH_3}{\underset{H}{C}}-CH_3$

(b) $H-\overset{O}{C}-\overset{CH_3}{\underset{CH_2CH_3}{C}}-CH_3$

(c) $CH_3CH_2-\underset{H}{C}-\overset{CH_3}{\overset{O}{C}}-CH_3$

(d) $CH_3-\overset{O}{C}-\overset{CH_3}{\underset{CH_3}{C}}-CH_3$

6. Which is the main final product of the following diazotization of an amine?

Reagents: $NaNO_2$, HCl, H_2O

(a)

(b)

(c)

(d)

7. **Which is the main product of the following reaction of an amide?**

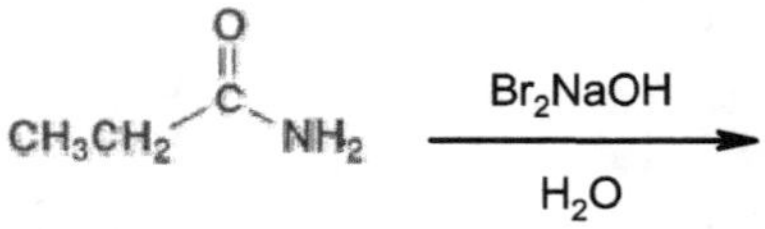

(a) $CH_3CH_2CH_2NH_2$

(b) $CH_3CH_2NH_2$

(c) 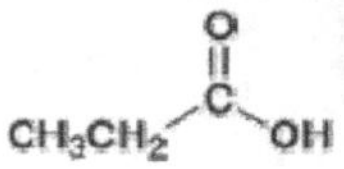 [structure: $CH_3CH_2C(=O)OH$]

(d) 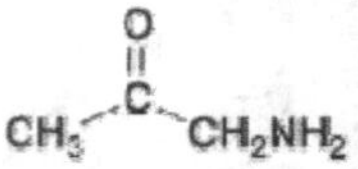 [structure: $CH_3C(=O)CH_2NH_2$]

[reaction: $CH_3CH_2C(=O)NH_2$ $\xrightarrow[H_2O]{Br_2,\ NaOH}$]

8. **Which of the following equations shows an unlikely main product?**

(a) 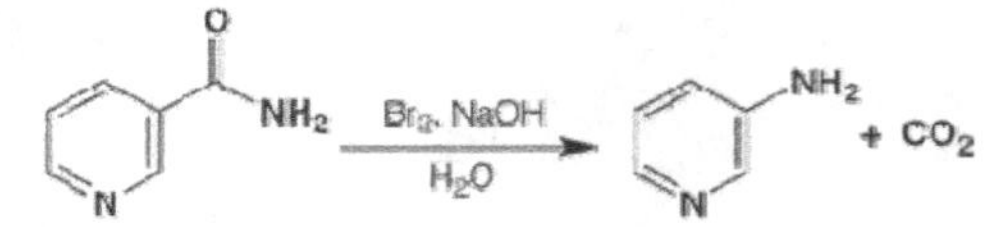 [pyridine-3-carboxamide $\xrightarrow[H_2O]{Br_2,\ NaOH}$ 3-aminopyridine + CO_2]

(b) 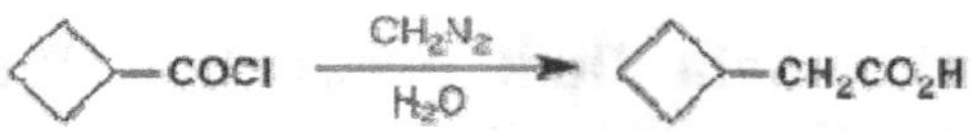 [cyclobutyl-COCl $\xrightarrow[H_2O]{CH_2N_2}$ cyclobutyl-CH_2CO_2H]

(c) 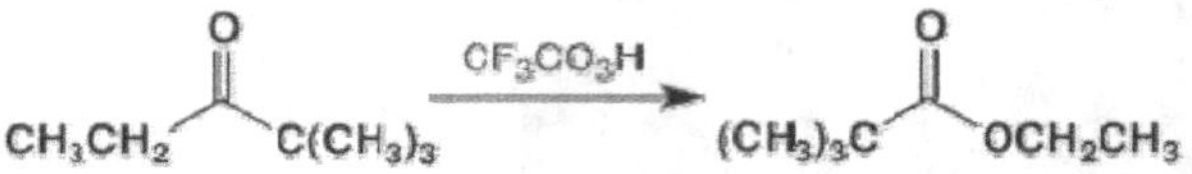 [$CH_3CH_2C(=O)C(CH_3)_3$ $\xrightarrow{CF_3CO_3H}$ $(CH_3)_3C\text{-}C(=O)OCH_2CH_3$]

(d) 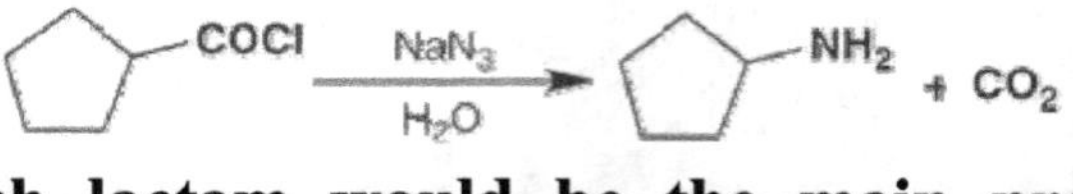 [cyclopentyl-COCl $\xrightarrow[H_2O]{NaN_3}$ cyclopentyl-NH_2 + CO_2]

9. **Which lactam would be the main product of the following rearrangement of an oxime when the *cis-trans* isomerization of the oxime is retarded?**

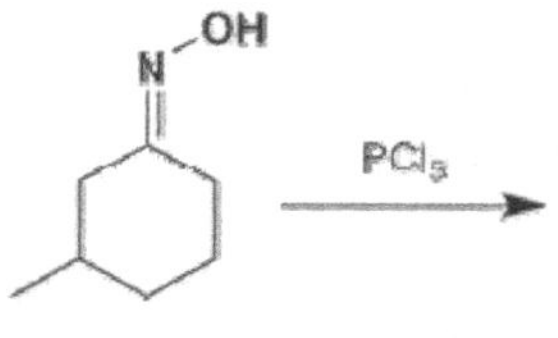

(a) 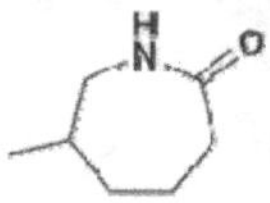 [seven-membered lactam structure]

(b) 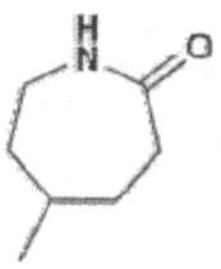 [seven-membered lactam structure]

[reaction: 4-methylcyclohexanone oxime $\xrightarrow{PCl_5}$]

(c) 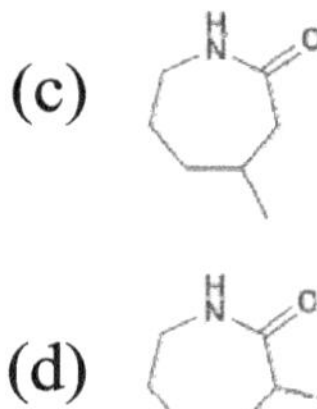

(d)

10. **Which of the following statements is wrong?**

    (a) Tertiary alkyl groups generally undergo 1,2-shifts to the electron-deficient centres more readily than primary and secondary alkyl groups do.

    (b) The bond breaking and making are concerted in the 1,2-shift of a methyl group in carbenium ions.

    (c) The transition structure for a 1,2-shift to an electron-deficient centre is regarded as a cyclic three-centre two-electron system.

    (d) The configuration of a carbon centre undergoing a 1,2-shift in a carbenium ion is not retained.

11. **Which alkene is the main product of acid-catalysed dehydration of the following alcohol? The reaction may involve a rearrangement.**

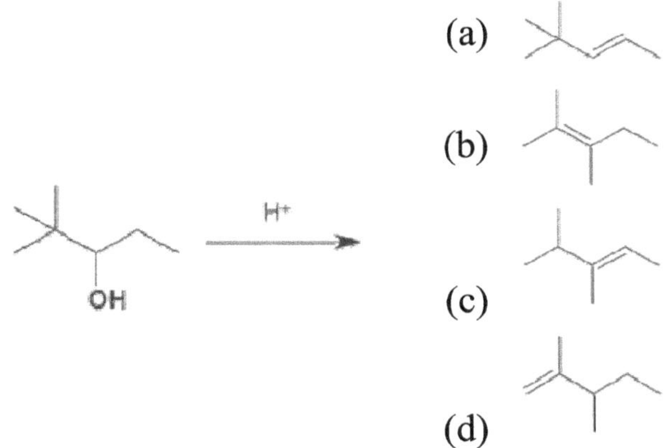

12. Which is the main product of the following rearrangement of a bicyclic diol?

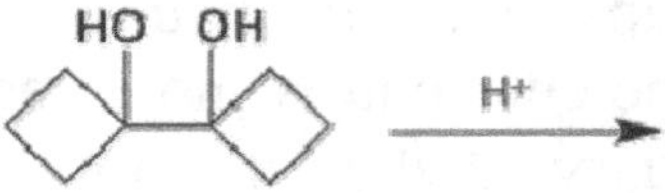

(a)

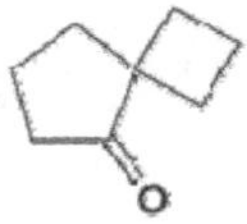

(b)

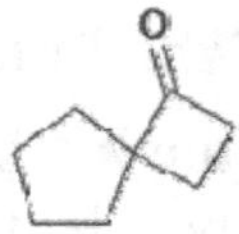

(c)

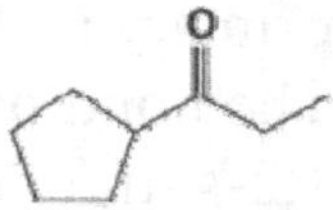

(d)

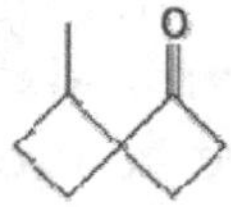

13. The following diol (1) and bromo alcohol (2) could possibly give carbonyl compound(s), A and/or B, by the indicated reactions. Which are the expected results?

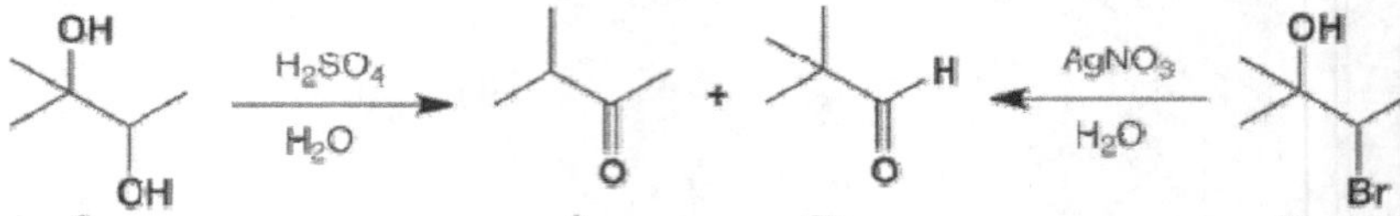

(a) Both 1 and 2 give mainly A.

(b) Both 1 and 2 give mainly B.

(c) Diol 1 gives A but 2 gives B as the main product.

(d) Diol 1 gives B but 2 gives A as the main product.

14. Diazotization of 1-aminomethyl-3-methyl cyclo pentanol may give 3-methyl- and 4-methyl cyclo hexanones as shown below. We represent the rebonding during the formation of 4-methyl cyclo hexanone as [1,2 ? 1,2]; this indicates that when the C1-C2 bond of the cyclo pentanol is broken, the C1-C2 bond of the cyclo hexanone is formed. How, in the same way, would we represent the bond breaking and making in the formation of the 3-methyl cyclohexanone?

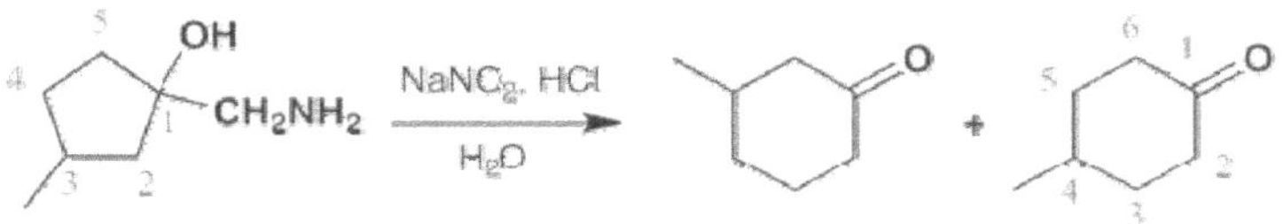

(a) $[1,2 \rightarrow 2,3]$ (b) $[1,2 \rightarrow 5,6]$

(c) $[1,5 \rightarrow 5,6]$ (d) $[1,5 \rightarrow 1,2]$

15. Which is an unlikely product in ethanolysis of the following methoxy tosylate?

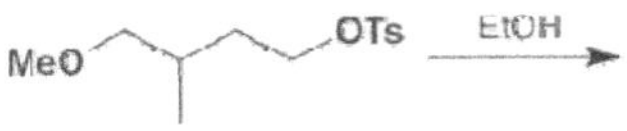

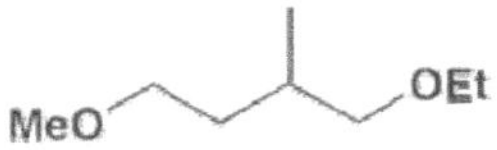

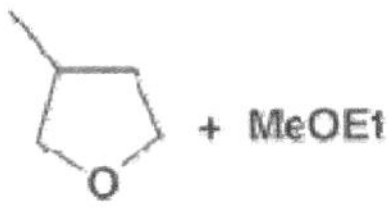

16. Which is an unlikely product of the following solvolysis?

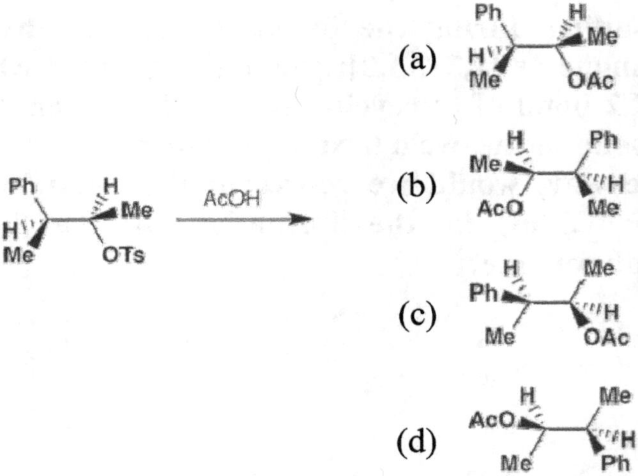

(a)

(b)

(c)

(d)

17. Which of the following reactions shows an unlikely result?

(a)

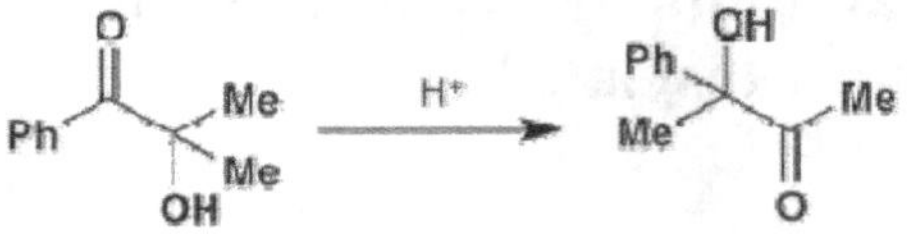

(b)

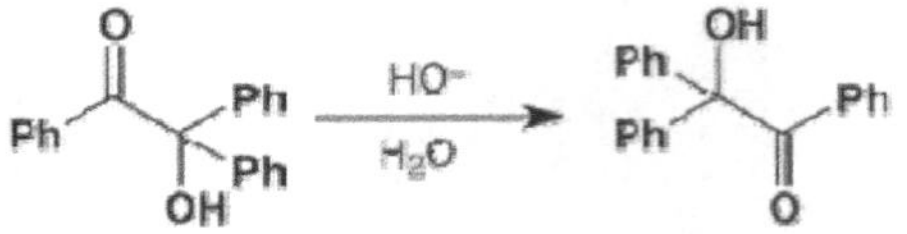

(c)

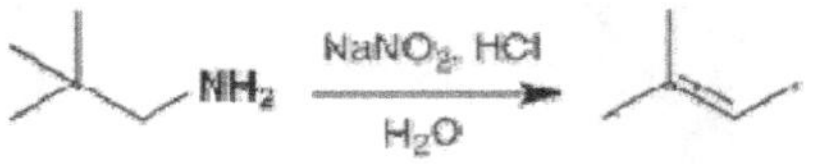

(d)

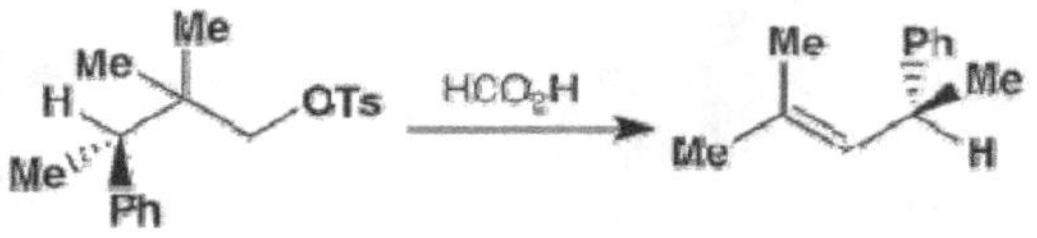

18. **Which is the main product of the following reaction of an amide?**

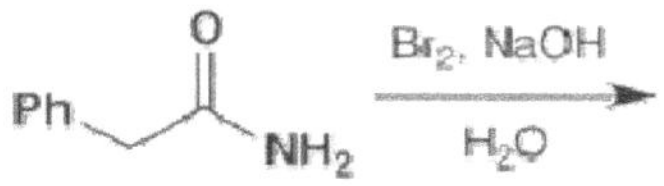

(Ph–CH$_2$–C(=O)–NH$_2$) $\xrightarrow[\text{H}_2\text{O}]{\text{Br}_2,\ \text{NaOH}}$

(a) Ph–CH$_2$–C(=O)–Br

(b) 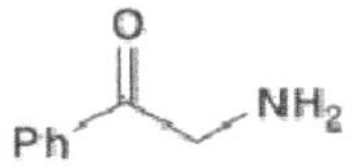 Ph–C(=O)–CH$_2$–NH$_2$

(c) Ph–CH$_2$–NH$_2$

(d)  Ph–CH$_2$–CH$_2$–NH$_2$

19. **Which is the main product of the following reaction of an acyl azide?**

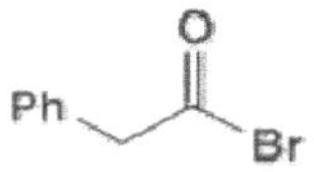

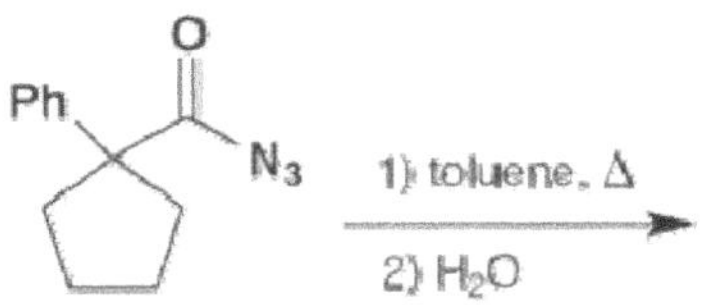

(1-phenylcyclopentyl)–C(=O)–N$_3$ $\xrightarrow[\text{2) H}_2\text{O}]{\text{1) toluene, }\Delta}$

(a) 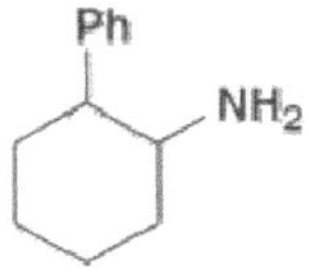 2-phenylcyclohexyl–NH$_2$ (1-phenyl, 2-amino cyclohexane)

(b) 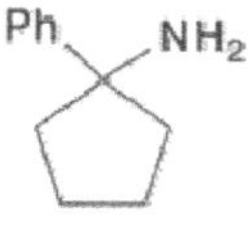 1-phenyl-1-amino-cyclopentane

(c) 1-phenylcyclopentane–C(=O)–OH

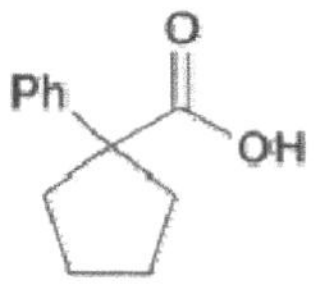

(d) 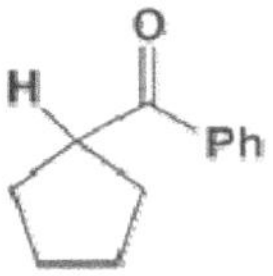 1-H-cyclopentane–C(=O)–Ph

20. **Which of the following reactions shows an unlikely main product?**

(a) OTs-bicyclic compound $\xrightarrow{\text{AcOH}}$ AcO-bicyclic compound

(b) $\underset{Ph}{\overset{Me\ \ Me}{\diagup}}$—Br $\xrightarrow[\text{$t$-BuOH}]{\text{$t$-BuOK}}$ $\underset{Me}{\overset{Me}{\diagup}}$=Ph

(c) $\underset{Ph}{\overset{O}{\diagdown}}$—$NH_2$ $\xrightarrow[\text{MeOH}]{\text{Br}_2,\ \text{NaOMe}}$ $PhNHCO_2Me$

(d) 2-chlorocyclohexanone $\xrightarrow[\text{MeOH}]{\text{NaOMe}}$ cyclopentane–CO_2Me

---

KEYS

1. (d)	**2.** (b)	**3.** (b)	**4.** (d)	**5.** (c)
6. (d)	**7.** (b)	**8.** (c)	**9.** (a)	**10.** (d)
11. (b)	**12.** (a)	**13.** (a)	**14.** (c)	**15.** (b)
16. (d)	**17.** (d)	**18.** (c)	**19.** (b)	**20.** (a)

1. Which reaction will not provide a synthesis of the following

(a) $CH_3CH_2CH_2$—MgBr + (ketone) $\xrightarrow{\text{1) Et}_2\text{O} \quad \text{2) H}_3\text{O}^+}$

(b) (isobutyl)—MgBr + (aldehyde)—H $\xrightarrow{\text{1) Et}_2\text{O} \quad \text{2) H}_3\text{O}^+}$

(c) (ester)—OMe + 2 CH_3MgI $\xrightarrow{\text{1) Et}_2\text{O} \quad \text{2) H}_3\text{O}^+}$

(d) (ketone) + CH_3MgI $\xrightarrow{\text{1) Et}_2\text{O} \quad \text{2) H}_3\text{O}^+}$

2. Which reaction is not appropriate for the synthesis of the following?

(a) (phenethyl)MgBr + CH_3CN $\xrightarrow{\text{1) Et}_2\text{O} \quad \text{2) H}_3\text{O}^+}$

(b) (phenyl)MgBr + (methyl vinyl ketone) $\xrightarrow{\text{1) Et}_2\text{O} \quad \text{2) H}_3\text{O}^+}$

(c) (benzyl)Br + (β-ketoester)OEt $\xrightarrow{\text{1) NaOEt, EtOH} \quad \text{2) H}_2\text{SO}_4, \text{H}_2\text{O}, \Delta}$

(d) (aryl)Li + (methyl vinyl ketone) $\xrightarrow{\text{1) THF} \quad \text{2) H}_3\text{O}^+}$

3. **Which of the following (to be converted by functional group interconversion, FGI) is not a good alternative target for the synthesis of pentan-2-ol?**

(a)

(b)

(c)

(d)

4. **Combination of reagents is wrong for disconnections (a)-(d) in the following?**

(a)

(b)

(c)

(d)

5. **Which of the protected forms of an alcohol ROH most readily gives back the alcohol by acidic hydrolysis?**

(a) RO—C(=O)Me

(b) RO—CMe$_3$

(c) RO—CH$_2$Ph

(d)

6. Which of the protected alcohols is unstable to an aqueous base?

(a) $RO\text{-}C(=O)\text{-}Me$

(b) $RO\text{-}CMe_3$

(c) $RO\text{-}CH_2\text{-}C_6H_5$

(d) (tetrahydropyranyl ether, $RO\text{-}$)

7. Which compound can be used for the synthesis of the following in a cyclization reaction?

(methyl 2-oxocyclohexanecarboxylate)

(a) $MeO_2C\text{-}CH_2CH_2\text{-}C(=O)\text{-}CH_3$

(b) $MeO_2C\text{-}CH_2CH_2CH_2\text{-}C(=O)\text{-}OMe$

(c) $MeO_2C\text{-}CH_2CH_2CH_2CH_2\text{-}C(=O)\text{-}CH_3$

(d) $MeO_2C\text{-}CH_2CH_2CH_2\text{-}C(=O)\text{-}OMe$

8. 4-Hydroxymethylcyclohexanone can be synthesized from a Diels-Alder adduct in the following reactions. Which combination of reagents is appropriate for the second step?

(a) 1. $NaBH_4$/MeOH;
   2. H_3O^+

(b) 1. $NaBH_4$/THF;
   2. $NaOH/H_2O$

(c) 1. $LiAlH_4/Et_2O$;
   2. H_3O^+

(d) 1. $LiAlH_4/Et_2O$;
   2. $NaOH/H_2O$

9. Which of (a)-(d) is the intermediate compound A in the following transformation?

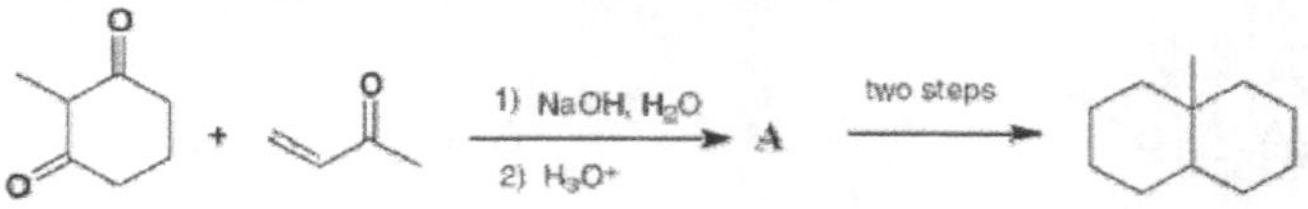
1) NaOH, H_2O
2) H_3O^+
A
two steps

(a) 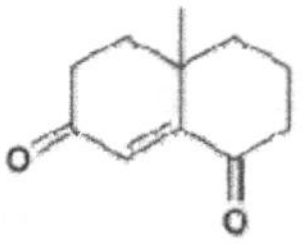

(b) 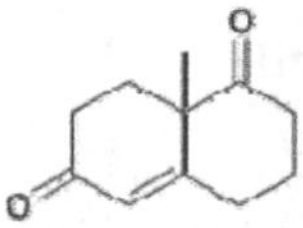

(c) 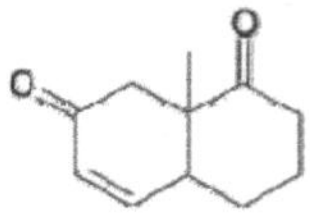

(d) 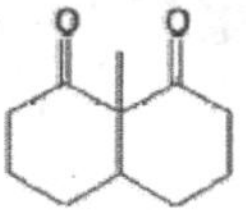

10. An α-amino acid, L-proline, can be used as a catalytic chiral auxiliary for a stereoselective aldol reaction. Which of (a)-(d) is not involved in the following transformation?

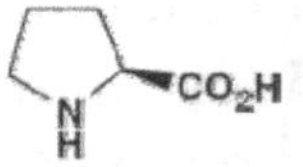
CO_2H
N
H

L-proline

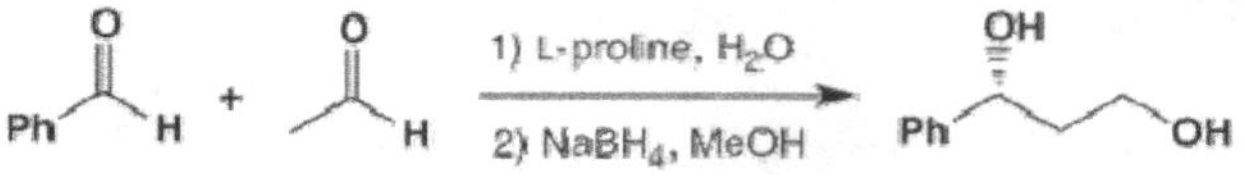
Ph
H
H
1) L-proline, H_2O
2) $NaBH_4$, MeOH
OH
Ph
OH

(a) 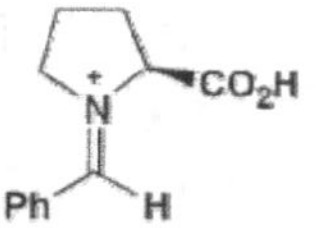

(b) 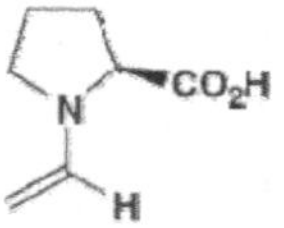

(c) 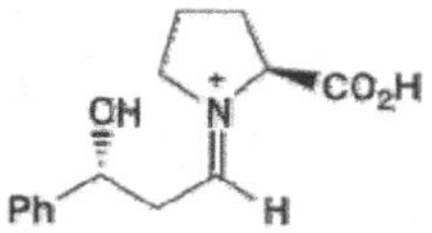

(d) 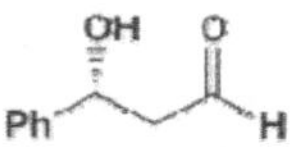

11. Which reaction will not provide a synthesis of 1,1-diphenylethanol?

(a) Ph–CO–Ph + MeMgBr 1) Et$_2$O 2) H$_3$O$^+$

(b) Ph–CO–Me + PhMgBr 1) Et$_2$O 2) H$_3$O$^+$

(c) Ph–CHO + PhCH$_2$MgBr 1) Et$_2$O 2) H$_3$O$^+$

(d) CH$_3$–CO–OEt + 2 PhMgBr 1) Et$_2$O 2) H$_3$O$^+$

12. Which reaction is not appropriate for the synthesis of pentan-2-one?

(a) CH$_3$CH$_2$CH$_2$MgBr + CH$_3$–CO–NMe$_2$ 1) Et$_2$O 2) H$_3$O$^+$

(b) CH$_3$CH$_2$CH$_2$Br + CH$_3$–CO–CH$_2$–CO–OEt 1) NaOEt, EtOH 2) H$_2$SO$_4$, H$_2$O, Δ

(c) CH$_3$CH$_2$CH$_2$C≡CH Hg(OAc)$_2$ / H$_2$SO$_4$, H$_2$O

(d) CH$_3$CH$_2$CH$_2$C≡CH 1) BH$_3$ / THF 2) H$_2$O$_2$ / NaOH

13. Which of the following (to be converted by functional group interconversion, FGI) is not a good alternative target for the synthesis of a primary amine, RCH$_2$NH$_2$?

(a) RCH$_2$Cl   (b) RCH$_2$OH   (c) RCHO   (d) RCN

14. **Which of the following (to be converted by FGI) is not a good alternative target for the synthesis of a carboxylic acid, RCO_2H?**

    (a) ROH   (b) RCHO   (c) $R-\overset{O}{\underset{\|}{C}}-CH_3$ (d) RCN

15. **A disconnection of 1,4-diphenylbut-2-yne is given below followed by possible corresponding synthetic reactions. Which of (a)-(d) does not include a suitable first reagent?**

    Disconnection:

    Synthetic steps for the disconnection shown:

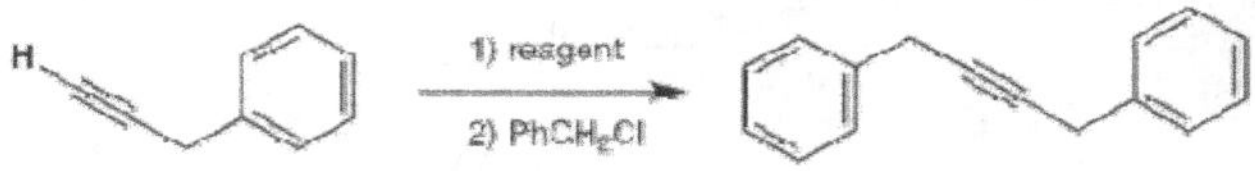

    (a) BuLi, THF   (b) EtMgBr, Et_2O

    (c) NaOEt, EtOH   (d) $NaNH_2$, NH_3

16. **Which of the protected forms of an amine R_2NH is relatively stable to an acid or a base in water?**

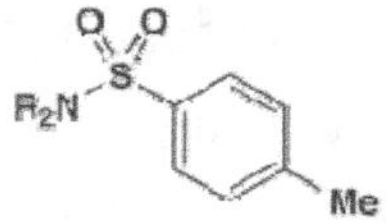

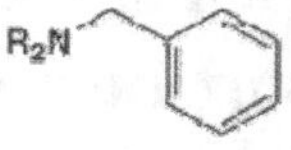

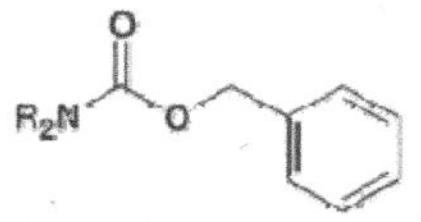

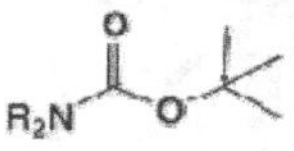

17. Which reaction conditions are not appropriate for the following transformation?

(a) Zn (Hg) / HCl

(b) H_2NNH_2 / NaOH

(c) $LiAlH_4$ / Et_2O

(d) $HSCH_2CH_2CH_2SH$ / H^+, then H_2 / Ni

18. Which combination of reagents is appropriate for the following transformation?

(a) 1) $HOCH_2CH_2OH$, H^+; 2) $LiAlH_4$, Et_2O; 3) H_3O^+

(b) 1) $NaBH_4$, MeOH; 2) $LiAlH_4$, Et_2O; 3) H_3O^+

(c) 1) $LiAlH_4$, Et_2O; 2) H_3O^+

(d) $NaBH_4$, MeOH

19. Which of (a)-(d) is the most suitable starting material for the synthesis of *m*-ethylaniline?

(a) 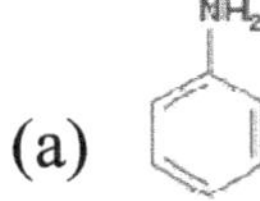 (b) 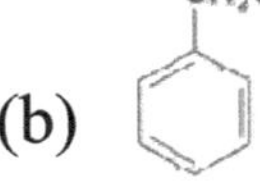 (c) 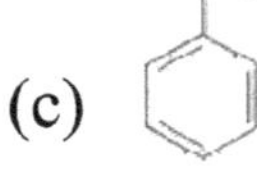 (d) 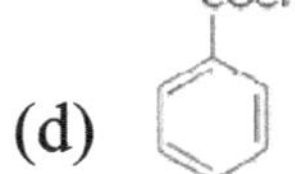

20. A secondary amine 1 (R_2NH) can be used as a catalytic chiral auxiliary for the conjugate addition of *N*-methylindole to but-2-enal. Which of (a)-(d) is

unlikely to be involved in the following "one pot" transformation?

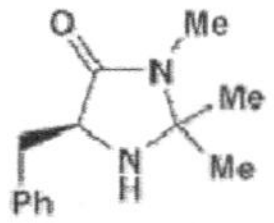

1 (R_2NH)

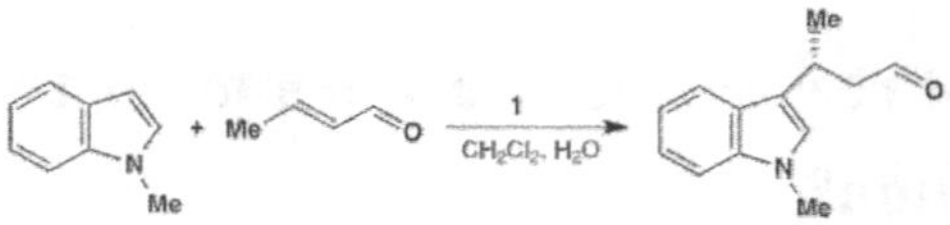

(a) 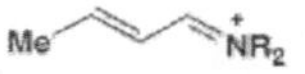

(b) 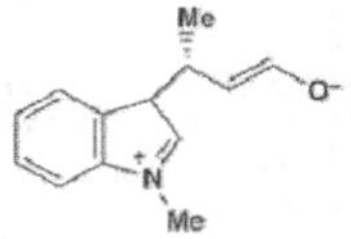

(c) 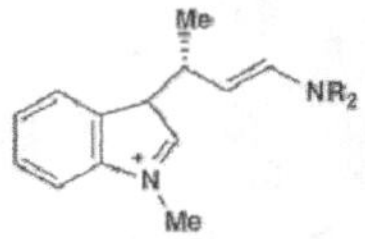

(d) 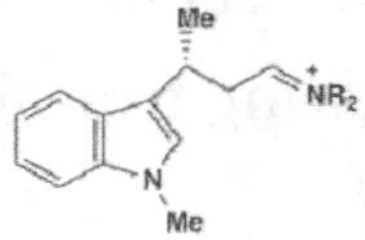

KEYS

1. (B)	**2.** (d)	**3.** (c)	**4.** (B)	**5.** (d)
6. (a)	**7.** (d)	**8.** (c)	**9.** (b)	**10.** (a)
11. (c)	**12.** (d)	**13.** (b)	**14.** (a)	**15.** (c)
16. (b)	**17.** (c)	**18.** (a)	**19.** (d)	**20.** (b)

CHEMISTRY OF BIOMOLECULES

1. **Which is the correct chair form of the β anomer of D-mannose?**

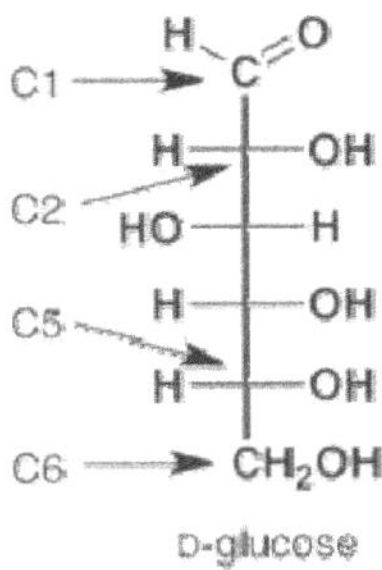

(a)

(b)

(c)

(d)

2. **Which carbon becomes the anomeric carbon of glucose in its pyranose form?**

(a) C1

(b) C2

(c) C5

(d) C6

3. **Which of the following statements regarding the reducing ability of a sugar is wrong?**

   (a) The aldehyde group of a saccharide is responsible for its reducing properties.

   (b) Ketoses are not reducing sugars because they are not aldehydes.

   (c) D-Glucose in predominantly in a cyclic hemiacetal form but it is a reducing sugar through the acyclic form with which the hemiacetal is in equilibrium.

   (d) A methyl glucoside is not a reducing sugar.

4. **The following are the four heteroaromatic bases found in DNA. Which base pair can form three hydrogen bonds?**

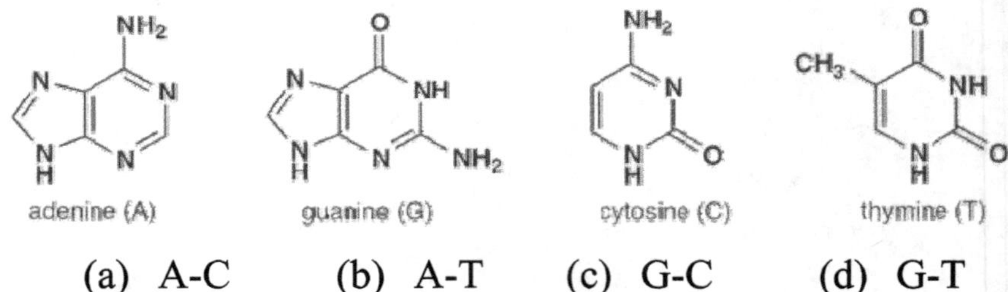

   (a) A-C   (b) A-T   (c) G-C   (d) G-T

5. **Which of (a)-(d) is complementary to the DNA segment 5'-ATGAGCCAT-3'?**

   (a) 5'-TACTCCGTA-3'   (b) 5'-TACTCGGTA-3'

   (c) 5'-TCATCGGTA-3'   (d) 5'-TACTGCGTA-3'

6. **Which naturally occurring α-amino acid is achiral?**

   (a) glycine   (b) glutamine   (c) leucine   (d) serine

7. Which structure is wrong for an L-α-amino acid?

(a) 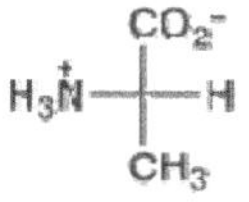 $H_3\overset{+}{N}$—C—H with CO_2^- above and CH_3 below

(b) $H_3\overset{+}{N}$—C—H with CO_2^- above and CH_2SH below

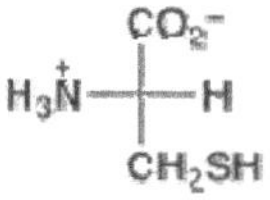

(c) structure with CO_2^-, H, CH_3 and $\overset{+}{N}H_3$

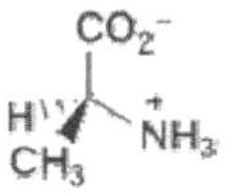

(d) 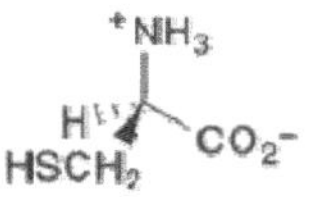 structure with $^+NH_3$, H, $HSCH_2$ and CO_2^-

8. The Edman degradation for the determination of peptide sequences involves a reaction using phenyl isothiocyanate to give a phenylthiohydantoin which identifies the *N*-terminal amino acid unit.

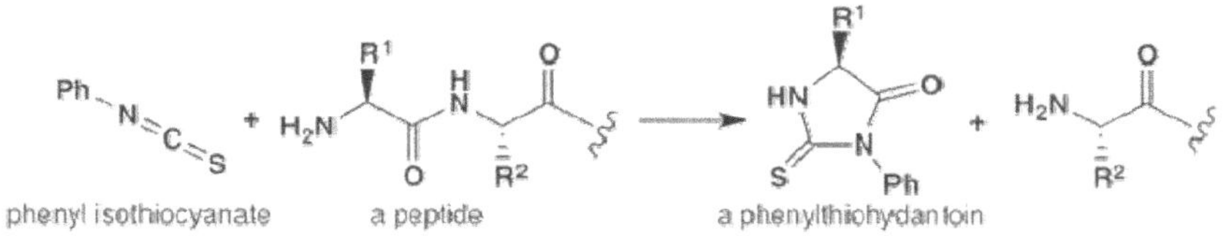

Ph—N=C=S + H₂N— (a peptide) → a phenylthiohydantoin + H₂N—

phenyl isothiocyanate   a peptide   a phenylthiohydantoin

Which of (a)-(d) shows a compound or partial structure which is not involved in reactions of the Edman degradation?

(a) 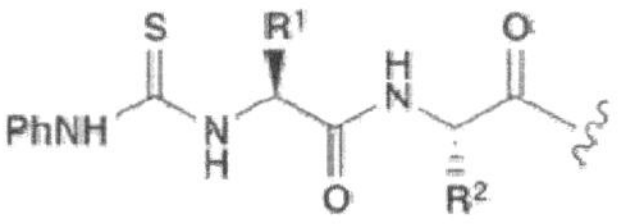 PhNH structure with S, R^1, O, R^2

(b) 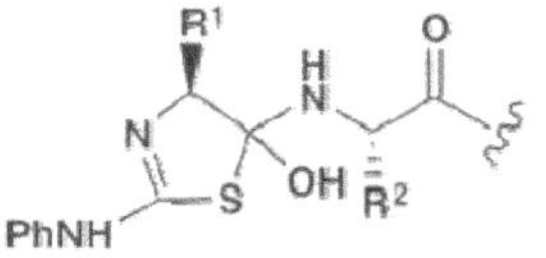 PhNH structure with R^1, N, S, OH, R^2, O

(c) 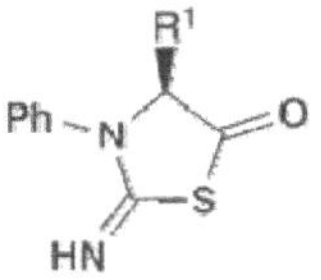 Ph—N structure with R^1, O, S, HN

(d) 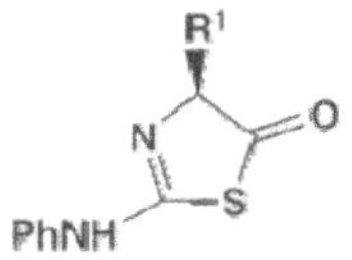 PhNH structure with R^1, N, O, S

9. The following is a structure of menthol without showing stereochemistry. How many chirality centres does menthol have and how many

stereoisomers are possible? Choose the correct combination of numbers.

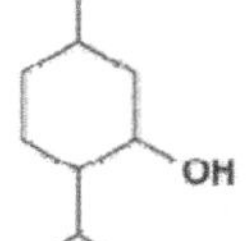

    (a) 3 centres and 9 isomers
    (b) 3 centres and 8 isomers
    (c) 3 centres and 6 isomers
    (d) 2 centres and 4 isomers

10. **Which of the following statements is wrong?**

(a) The configuration of L-serine is S, but that of L-cysteine is R.

(b) Saturated fatty acids have generally lower melting points than the unsaturated fatty acids.

c) The anions of fatty acid salts (soaps) have a hydrophilic polar end and a hydrophobic hydrocarbon chain, while phospholipids have two hydrophobic chains.

d) A sesquiterpene is an unsaturated hydrocarbon with 15 carbon atoms.

11. **Which is the correct chair form of the α anomer of D-galactose?**

12. **Which is the correct assignment of configurations of chirality centres, C2-C5, of D-glucose?**

(a) $2R, 3S, 4R, 5R$

(b) $2S, 3R, 4S, 5S$

(c) $2R, 3R, 4R, 5R$

(d) $2S, 3S, 4R, 5R$

13. **Which of the following is not a reducing sugar?**

(a) D-fructose

(b) D-ribose

(c) cellobiose

(d) sucrose

14. **Which of the following monosaccharides gives an optically inactive product upon oxidation with HNO_3?**

(a) D-glucose

(b) D-mannose

(c) D-galactose

(d) D-fructose

15. **How many chirality centres does S-adenosylme thionine (SAM) have?**

(a) 3

(b) 4

(c) 5

(d) 6

16. **Which of (a)-(d) is complementary to the DNA segment 5'-ACGTAATC-3'?**

(a) 5'-TGCATTCG-3'

(b) 5'-TGCATTAG-3'

(c) 5'-TGCATAAG-3'

(d) 5'-TGACTTAG-3'

17. Which amino acid has the lowest isoelectric point?

(a) glycine

(b) serine

(c) glutamic acid

(d) lysine

18. Which of the following amino acids has the highest isoelectric point?

(a) glycine

(b) serine

(c) glutamic acid

(d) lysine

19. Which of (a)-(d) correctly indicates the configurations of the two chirality centres, C2 and C3, of sphingomyelin?

sphingomyelin

(a) $2S,3R$

(b) $2R,3S$

(c) $2S,3S$

(d) $2R,3R$

20. Which of the following statements is false?

(a) Natural fatty acids contain even numbers of carbon atoms.

(b) Diterpenes contain 10 carbons.

(c) Eicosanoids have structures derived from arachido nic acid.

(d) Arachidonic acid is a C_{20} unsaturated carboxylic acid.

KEYS

1. (d)	**2.** (a)	**3.** (b)	**4.** (c)	**5.** (b)
6. (a)	**7.** (d)	**8.** (c)	**9.** (b)	**10.** (b)
11. (b)	**12.** (a)	**13.** (d)	**14.** (c)	**15.** (c)
16. (b)	**17.** (c)	**18.** (d)	**19.** (a)	**20.** (b)

1. **Which of the following statements is wrong?**

   (a) UV absorption is attributable to electronic transitions.

   (b) UV spectra provide information about valence electrons.

   (c) IR absorption is attributable to transitions between rotational energy levels of whole   molecules.

   (d) NMR spectrometers use radiofrequency electro magnetic radiation.

2. **Which is the correct order of increasing wave number of the stretching vibrations of (1) C-H (alkane), (2) O-H (alcohol), (3) C=O (ketone), and (4) C≡C (alkyne)?**

   (a) $(4) < (3) < (2) < (1)$ (b) $(3) < (4) < (2) < (1)$

   (c) $(3) < (4) < (1) < (2)$ (d) $(4) < (3) < (1) < (2)$

3. **How many signals does the aldehyde $(CH_3)_3CCH_2CHO$ have in ^1H NMR and ^{13}C NMR**

spectra?

(a) five ^{1}H signals and six ^{13}C signals

(b) three ^{1}H signals and four ^{13}C signals

(c) five ^{1}H signals and four ^{13}C signals

(d) three ^{1}H signals and six ^{13}C signals

4. **Which of hydrogens (a)-(d) in the following molecule gives a triplet signal in a normal ^{1}H NMR spectrum?**

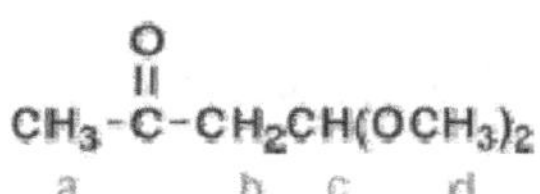

(a) hydrogen a

(b) hydrogen b

(c) hydrogen c

(d) hydrogen d

5. **Which hydrogen of 1-chloropent-2-ene shows the largest chemical (downfiel(d) shift in its NMR spectrum?**

(a) the H on C1

(b) the H on either C2 or C3

(c) the H on C4

(d) the H on C5

6. **Which carbon of (a)-(d) of hex-3-en-2-one shows the largest (most downfiel(d) chemical shift in the NMR spectrum?**

(a) C1   (b) C2   (c) C4   (d) C6

7. **Which of the following statements regarding IR spectroscopy is wrong?**

   (a) Infrared radiation is higher in energy than UV radiation.

   (b) Infrared spectra record the transmission of IR radiation.

   (c) Molecular vibrations are due to periodic motions of atoms in molecules, and include bond stretching, torsional changes, and bond angle changes.

   (d) Infrared spectra give information about bonding features and functional groups in molecules.

8. **Which of the following statements regarding NMR spectroscopy is wrong?**

   (a) NMR signals towards the left of the spectral chart correspond to larger chemical shifts.

   (b) Chemical shifts are larger when the frequencies of the radiation which induces the nuclear transitions are higher.

   (c) Chemical shifts are larger when shielding effects are greater.

   (d) A hydrogen signal splits into n+1 peaks by spin-spin coupling when the number of equivalent hydrogen atoms on adjacent atom(s) is n, and no other neighbouring atoms are involved.

9. **Which of the following statements regarding mass spectrometry is wrong?**

   (a) In a normal mass spectrometer, electron impact causes a molecule to lose an electron and become a molecular radical cation which decomposes into fragment cations and radicals.

(b) Only cations can be detected by a normal mass spectrometer.

(c) A compound whose molecules contain just one bromine atom shows two molecular ion peaks of similar intensity, one at +1 and one at -1 of the average m/z value.

(d) Molecular ion peaks always have even-numbered values of m/z.

10. Which of the following statements is wrong?

(a) A conventional mass spectrometer employs high energy UV radiation.

(b) A conventional mass spectrometer does not employ a spectrophotometric detector.

(c) Conventional mass spectrometry does not always require samples of high purity.

(d) A mass spectrum does not show signals due to uncharged radicals.

11. Absorption of radiation in the UV range attributable to n→π* electronic transitions is characteristic of which of the following types of compounds?

(a) Aromatic hydrocarbons.

(b) Unsaturated carbonyl compounds.

(c) Non-conjugated polyenes.

(d) Conjugated polyenes.

12. Which of the following statements is wrong.

(a) The wavenumber of a band in an IR spectrum is proportional to the frequency of the associated molecular vibration.

(b) Water is a good solvent for recording UV spectra of water-soluble compounds.

(c) Water is a good solvent for recording IR spectra of water-soluble compounds.

(d) Hydrogen bonding in hydroxy compounds leads to broadening of spectral bands attributable to O-H stretching vibrations.

13. **Which is the correct order of increasing wave number of the stretching vibrations of (1) C-H (alkane), (2) C-H (alkene), (3) C-H (alkyne), and (4) C-H (arene)?**

   (a) $(1) < (2) \approx (3) < (4)$ (b) $(4) < (3) \approx (2) < (1)$

   (c) $(3) < (4) \approx (2) < (1)$ (d) $(1) < (4) \approx (2) < (3)$

14. **How many signals does the unsaturated ketone $(CH_3)_2CHCH_2C(O)CH=CH_2$ have in ^1H NMR and ^{13}C NMR spectra?**

   (a) five ^1H signals and six ^{13}C signals
   (b) six ^1H signals and six ^{13}C signals
   (c) six ^1H signals and seven ^{13}C signals
   (d) five ^1H signals and seven ^{13}C signals

15. **Which of the following statements in the context of ^1H NMR spectroscopy is true?**

   (a) Arene C-H chemical shift (δ) values are greater than simple alkenes C-H chemical shift values because of the aromatic ring current.

   (b) Arene C-H chemical shift (δ) values are smaller than simple alkenes C-H chemical shift values be cause of the aromatic ring current.

(c) Arene C-H signals are always multiplets.

(d) Arene C-H signals are always singlets.

16. **Which carbon of (a)-(d) of hex-3-en-2-one has the smallest (most upfiel(d) chemical shift in the NMR spectrum?**

(a) C1   (b) C2   (c) C4   (d) C6

17. **Which of (a)-(d) indicates the multiplicities for hydrogens on C1, C3, and C4 of butanone attributable to spin-spin coupling in its ^{1}H NMR spectrum.**

(a) Hs on C1, singlet; Hs on C3, doublet; Hs on C4, triplet.

(b) Hs on C1, singlet; Hs on C3, triplet; Hs on C4, quartet.

(c) Hs on C1, singlet; Hs on C3, quartet; Hs on C4, triplet.

(d) Hs on C1, triplet; Hs on C3, doublet; Hs on C4, triplet.

18. **Which of (a)-(d) indicates the correct order of carbon chemical shifts of the four carbons of the following compound.**

$$CH_2{=}CH{-}\overset{\overset{\textstyle O}{\|}}{C}{-}O{-}CH_3$$
$$\begin{matrix} C3 & C2 & C1 & C_{Me} \end{matrix}$$

(a) $C_{Me} < C2 < C3 < C1$

(b) $C_{Me} < C3 < C2 < C1$

(c) $C_{Me} < C2 < C1 < C3$

(d) $C_{Me} < C1 < C2 < C3$

19. Which of the following statements regarding electron-impact mass spectrometry is true?

(a) Samples need isotopic labels.

(b) The base peak is formed by loss of one electron from each vaporised molecule by an electron beam.

(c) Compounds must have a functional group to show a mass spectrum.

(d) A meaningful mass spectrum can sometimes be obtained on a very small sample of an impure compound.

20. Which of the following statements regarding mass spectrometry is false?

(a) The base peak of a simple ketone is usually attributable to an acylium ion.

(b) The molecular ion of carbonyl compounds with a - C-H readily undergoes elimination of an alkene to give a relatively stable enol radical cation.

(c) The molecular ion peak of some alcohols is very weak because it readily loses an alkyl radical to give a relatively stable oxonium (hydroxyl carbenium) ion.

(d) Structurally isomeric alkanes cannot be distinguished by low resolution mass spectrometry.

KEYS

1. (c)	**2.** (c)	**3.** (b)	**4.** (c)	**5.** (b)
6. (b)	**7.** (a)	**8.** (c)	**9.** (d)	**10.** (a)
11. (b)	**12.** (c)	**13.** (d)	**14.** (b)	**15.** (a)
16. (d)	**17.** (c)	**18.** (a)	**19.** (d)	**20.** (d)

CHAPTER 21

STEREOCHEMISTRY

1. **Which of the following groups has the highest priority according to the Cahn-Ingold-Prelog sequence rules?**

   (a) CH_3 (b) CH_2Cl (c) CH_2OH (d) CHO

2. **Which of the following groups has the highest priority according to the Cahn-Ingold-Prelog sequence rules?**

   (a) C≡CH (b) $CH{=}CH_2$

   (c) $CH(OH)CH_3$ (d) CH_2CH_2OH

3. **Which of the following groups has the lowest priority according to the Cahn-Ingold-Prelog sequence rules?**

   (a) C≡CH (b) $CH{=}CH_2$

   (c) $CH(OH)CH_3$ (d) CH_2CH_2OH

4. **Which of the following has the (*R*) configuration?**

   (a) 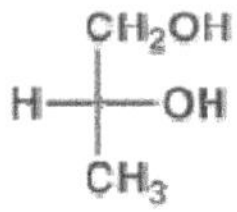

   (b) 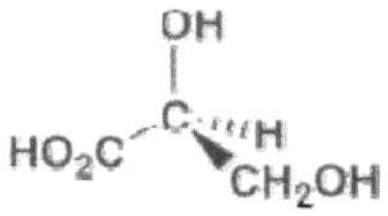

   (c) 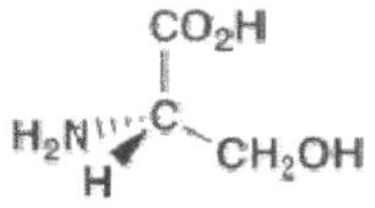

   (d) 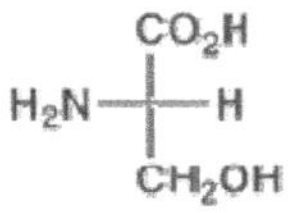

5. Which of (a)-(d) shows the same compound as the following?

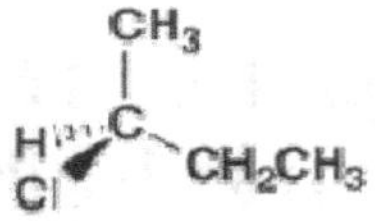

(a) 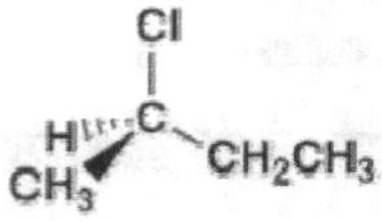

(b) 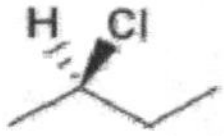

(c) 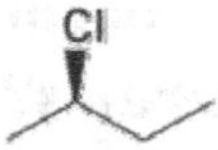

(d) 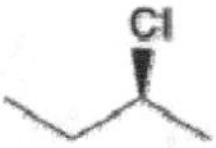

6. Which shows the same compound as the following?

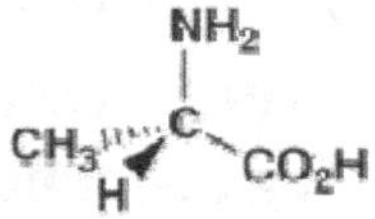

(a) 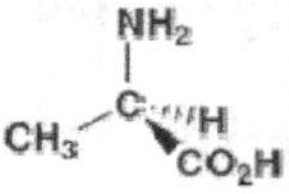

(b) 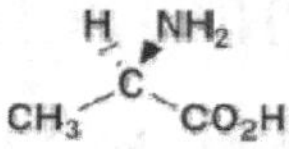

(c) 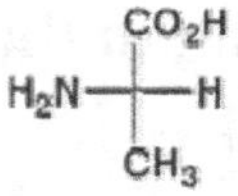

(d) 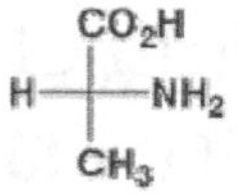

7. Which compound is the enantiomer of the following?

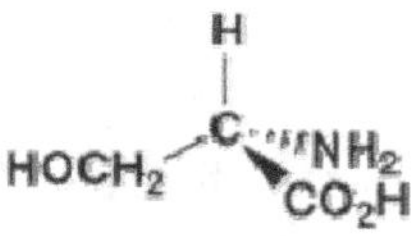

(a) 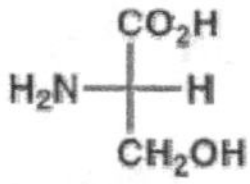

(b) 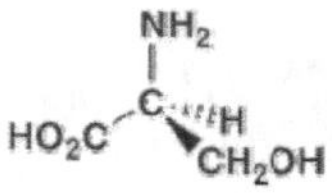

(c) 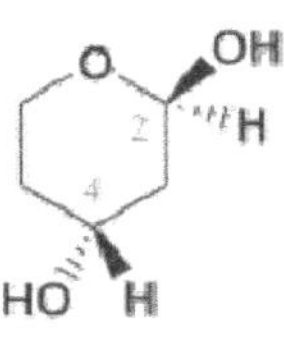

(d)

8. Which is the correct assignment of chirality at C2 and C4 of the following molecule?

(a) 2*S*,4*S*

(b) 2*R*,4*R*

(c) 2*S*,4*R*

(d) 2*R*,4*S*

9. Which is the correct assignment of chirality at C2 and C3 of the following molecule?

(a) 2*S*,3*S*

(b) 2*R*,3*R*

(c) 2*S*,3*R*

(d) 2*R*,3*S*

10. Which of the following compounds is achiral?

(a)

(b)

(c)

(d)

11. Which of the following compounds is achiral?

(a)

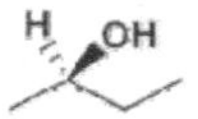

(b)

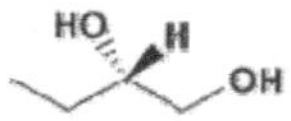

(c)

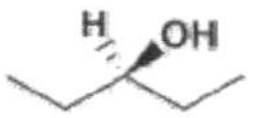

(d)

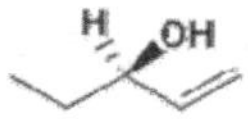

12. Which of the following groups has the highest priority in the Cahn-Ingold-Prelog sequence rules?

(a) CH_2CH_3

(b) $CH=CH_2$

(c) $C\equiv CH$

(d) $C(CH_3)_3$

13. Which of the following groups has the highest priority in the Cahn-Ingold-Prelog sequence rules?

(a) CH_2OH

(b) CH_2OCH_3

(c) $CH=O$

(d) CO_2H

14. Which of the following groups has the second highest priority in the Cahn-Ingold-Prelog sequence rules?

(a) CH_2OH

(b) CH_2OCH_3

(c) $CH=O$

(d) CO_2H

15. Which of the following has the (*R*) configuration?

(a)

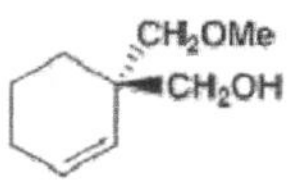

(b)

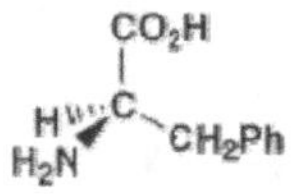

(c)

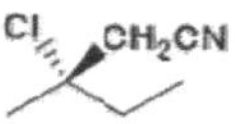

(d)

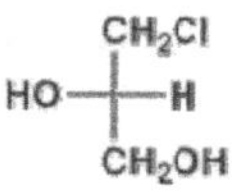

16. Which of the following has the (*S*) configuration?

(a)

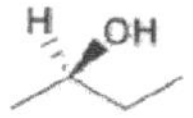

(b)

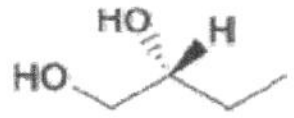

(c)

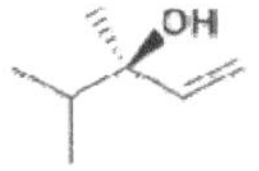

(d)

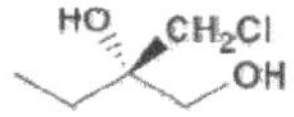

17. Which is the enantiomer of the following molecule?

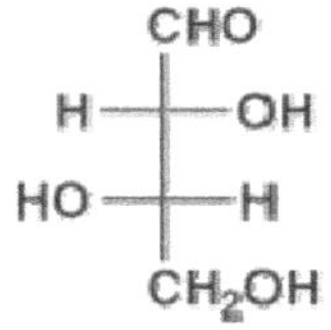

(a)

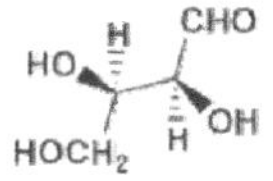

(b)

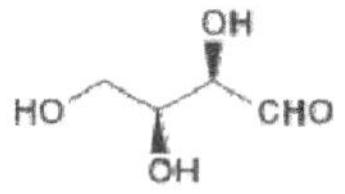

(c)

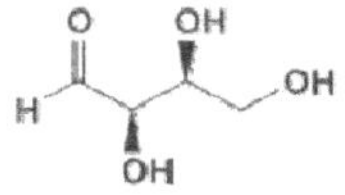

(d)

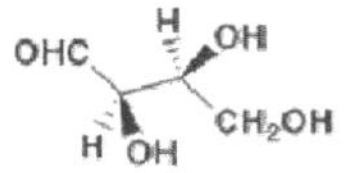

18. Which is the *meso* isomer of butane-1,2,3,4-tetraol?

(a)

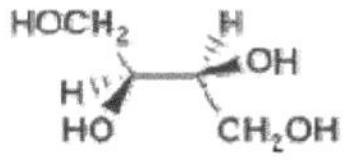

(b)

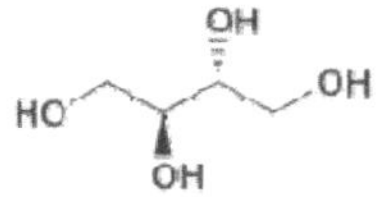

(c)

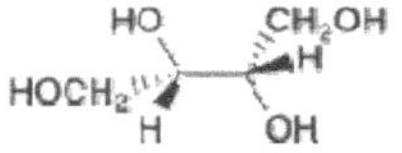

(d)

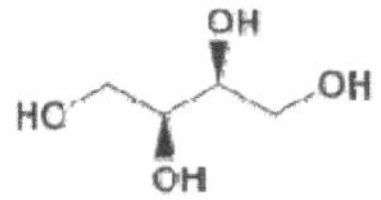

19. Which is the correct assignment of chirality at C1 and C4 of the following molecule?

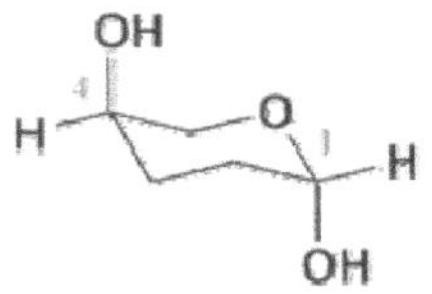

(a) $1S, 4R$

(b) $1R, 4R$

(c) $1R, 4S$

(d) $1S, 4S$

20. Which of the following reactions gives no chiral product?

(a) 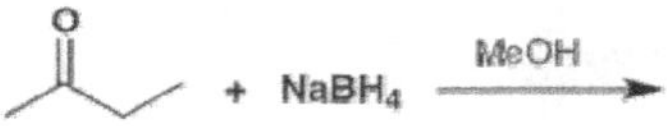 (propanal) + NaBH$_4$ $\xrightarrow{\text{MeOH}}$

(b) 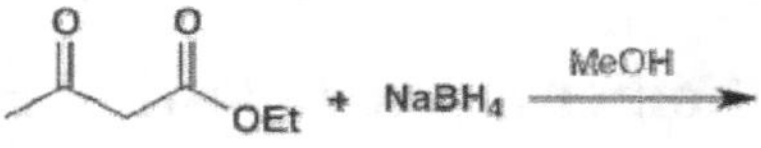 (ethyl acetoacetate) + NaBH$_4$ $\xrightarrow{\text{MeOH}}$

(c) 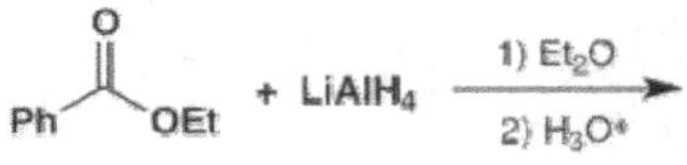 Ph–CO–OEt + LiAlH$_4$ $\xrightarrow[\text{2) H}_3\text{O}^+]{\text{1) Et}_2\text{O}}$

(d)  (β-ketoester, OEt) + LiAlH$_4$ $\xrightarrow[\text{2) H}_3\text{O}^+]{\text{1) Et}_2\text{O}}$

KEYS

1. (b)	**2.** (c)	**3.** (d)	**4.** (a)	**5.** (d)
6. (c)	**7.** (c)	**8.** (b)	**9.** (d)	**10.** (b)
11. (c)	**12.** (c)	**13.** (d)	**14.** (c)	**15.** (b)
16. (d)	**17.** (a)	**18.** (b)	**19.** (a)	**20.** (c)

BIBLIOGRAPHY

1. Advanced Organic Chemistry by Francis A. Carey, Richard A. Sundberg, 5th Edition, 200 7, ISBN-13: 978-0-387-44897-8, Springer.

2. An update of multiple ch oice questions (MCQs): the sequential multiple choice quest ions by V. Carta, A. Catalano, M. Ferappi, G. Lentini, F. Palluotto, P. Tortorella & V. Tortorella, Pharmacy Education, January 2009; 9 (1).

3. Conceptual Problems in Organic Ch emistry, by DK Singh, Pearson Publication, 2013.

4. March's Adv anced Organi c Che mistry. Reactions, Mechanisms, and Structure by Michael B. S mith, Jerry March, 6 th Edition, 2007, ISBN: 978-0- 471-72091-1, Wiley.

5. Modern Organic Sy nthesis - An Introduction b y George S. Z weifel, Mich ael H. Nantz, 1st Edition, 2007, ISBN: 978-0-716-77266-8, W. H. Freeman.

6. Organic Chemistry b y Jonathan Clayden, Nick Geeves, Stuart Warren, Second E   dition, 2012, ISBN: 978-0199270293, Oxford University Press.

7. Organic Chemistry b y Robert Thornto n Morrison, Robert Neilson Boyd,   Saibal Kanti seventh Edition, Pearson Publication.

8. Organic Mechanisms - Reactions, Stereochemistry and Synthesis by Reinhard Brückner,

9. Pericyclic Reactions - A Textbook, Sethuraman Sankararaman, First Edition, 2005, ISBN: 3-527-31439-3, Wiley-VCH, Weinheim.

10. Structure and Reactivity in Organic Chemistry by Mark G. Moloney, First Edition, 2008, ISBN: 978-1-4051-1451-6, Wiley-Blackwell.